ROBOTICS: An Introduction

ROBOTICS:
An Introduction
Second Edition

Douglas R. Malcolm, Jr.
GMFanuc, Inc.

DELMAR PUBLISHERS INC.®

For information, address Delmar Publishers Inc.
3 Columbia Circle, PO Box 15015
Albany, New York 12212-5015

Printed in the United States of America
Published simultaneously in Canada
by Nelson Canada
A Division of International Thomson Limited

ISBN 0-8273-3913-5

To my children
Pamela, Douglas, and Robert

Contents

CHAPTER 8

Servo System Control   184

CHAPTER 9

Robotic Gears and Linkages   207

≡ Preface ≡

The second edition of *Robotics: An Introduction* is a state-of-the-art textbook that introduces the technical student to flexible automation. This text is intended for students who will install, repair, or develop robotic flexible automation systems in the industrial environment. It is intended for use in engineering technology programs at the post-secondary level, but is also suitable for industrial technology programs curricula, as well as for in-service industrial training programs. Industrial managers, applications engineers, and production personnel who desire to become familiar with robotics or to introduce it into their production facilities will find this textbook useful.

This text requires that the student have a basic background of electronic circuits and devices, as well as a familiarity with digital logic circuits and the fundamental understanding of microprocessor technology. While not essential, an understanding of the principles of mechanics and fluid power technology is also helpful.

The presentation incorporates a variety of robotic systems that are found on the factory floor. The real-world examples provide literacy in the basic terminology of the subject and illustrate contemporary applications. It is assumed that many readers, after completing this text, will pursue further studies in robotics.

An introduction to basic robotics terms and components is presented in Chapter 1. The various levels of technology within the robotics industry are highlighted in this chapter. Chapter 2 identifies and defines the geometries associated with robots and describes the basic features of different manipulators found in industrial application. Chapter 3 explores the relationship between the robot controller and the manipulator, as well as the electronic circuits of the robot controller. Chapter 4 introduces the basic structure of robot command data and the programming structures that are found in various robots.

Chapter 5 describes those components that are operational aids for the robot—components such as teach pendants, operator's panels, and peripheral programming devices. Chapter 6 introduces the hydraulic and pneumatic drive systems used to power the robot manipulator, while Chapter 7 explains the electric drives that are used. Servo system control of the various degrees of freedom found on the manipulator is described in Chapter 8.

The basic concepts of the mechanics of the manipulator are explained in Chapter 9. Mechanical linkages used to convert the drive power into manipulator motion are discussed, along with the various styles of interfacing between peripheral devices and robotics systems, in Chapter 10.

The devices known as end effectors are described in Chapter 11. Sensors, such as the vision and tactile type, are discussed in Chapter 12. Applications taken from real-world installations are described in Chapter 13. These applications deal with material handling, machine load/unloading, die casting, welding, painting, and electronic assembly. Chapter 14 deals with the social–economic impact that robotics will have on U.S. industry and what U.S. workers will have to do to prepare for future automation impacts. Chapter 15 identifies the fundamentals of communication that is established between a robot controller and peripheral devices within the work cell.

The author gratefully acknowledges the input received from many robotics manufacturers, colleagues in the robotics industry, and instructors of robotics courses. Thanks are due to numerous individuals and firms for providing up-to-date technical information and photographs and, in many cases, for granting permission to use copyrighted material in this book. Appreciation is extended to the following companies: ASEA Robotics, Inc.; Automatix, Inc.; GMF Robotics, Inc.; Hobart Brothers Company; and the DeVilbiss Company.

Sincere thanks are offered to those robotics instructors who reviewed the second edition manuscript and offered suggestions, many of which were incorporated into the final draft: Professors Duane Olson, Southwestern Vocational–Technical Institute, Granite Falls, Minnesota; Lyle R. Langlois, Glendale Community College, Phoenix, Arizona; A.F. Adkins, Amarillo College, Amarillo, Texas; Ralph R. Stockard, Technical College of Alamance, Haw River, North Carolina; Don E. Holzhei, Delta College, University Center, Michigan; William L. Meier, Waukesha County Technical Institute, Pewaukee, Wisconsin; and C. Eddie Gore, Carroll Technical Institute, Carrollton, Georgia.

Finally, I extend my heartfelt thanks to my children Pam, Doug, and Rob for understanding when their father had to spend a great deal of time on this project. And last, but not least, I thank my wife Ruth for all her help. Without my family's encouragement and love and support, this second edition could have never been written.

ROBOTICS: An Introduction

1

Introduction to Robotics

OBJECTIVES

Upon completing this chapter, you should be familiar with:

— Robot terminology,
— Basic components of robots,
— Robot motion,
— Robot technology levels.

INTRODUCTION

The industrial robot and its operation is the topic of this text. The industrial robot is a tool that is used in the manufacturing environment to increase productivity. It can be used to do routine and tedious assembly line jobs, or it can perform jobs that might be hazardous to the human worker. For example, one of the first industrial robots was used to replace the nuclear fuel rods in nuclear power plants. A human doing this job might be exposed to harmful amounts of radiation. The industrial robot can also operate on the assembly line, putting together small components, such as placing electronic components on a printed circuit board. Thus, the human worker can be relieved of the routine operation of this tedious task. Robots can also be programmed to defuse bombs, to serve the handicapped, and to perform functions in numerous applications in our society.

The robot can be thought of as a machine that will move an end-of-arm tool, sensor, and/or gripper to a preprogrammed location. When the robot arrives at this location, it will perform some sort of task. This task could be welding, sealing, machine loading, machine unloading, or a host of assembly jobs. Generally, this work can be accomplished without the involvement of a human being, except for programming and for turning the system on and off.

In this chapter, you will become familiar with the various

components used to establish a robotic system. As you will see, robotic systems have different levels of technology that enable the robot to be a very simple operating machine or a complex, flexible, automated manufacturing cell.

ROBOT TERMINOLOGY

In this section and in the following sections, the basic terminology of robotic systems is introduced. Also, the fundamental components of a robotic system are identified and defined. Succeeding chapters of the text will expand on the basic terminology and definitions, giving more detailed information.

A **robot** is a reprogrammable, multifunctional manipulator designed to move parts, materials, tools, or special devices through variable programmed motions for the performance of a variety of different tasks. This basic definition leads to other definitions, presented in the following paragraphs, that give a complete picture of a robotic system.

Preprogrammed locations are paths that the robot must follow to accomplish work. At some of these locations, the robot will stop and perform some operation, such as assembly of parts, spray painting, or welding. These preprogrammed locations are stored in the robot's memory and are recalled later for continuous operation. Furthermore, these preprogrammed locations, as well as other program data, can be changed later as the work requirements change. Thus, with regard to this programming feature, an industrial robot is very much like a computer, where data can be stored and later recalled and edited.

The **manipulator** is the arm of the robot. It allows the robot to bend, reach, and twist. This movement is provided by the manipulator's **axes**, also called the **degrees of freedom** of the robot. A robot can have from 3 to 16 axes. In the remainder of this text, the term *degrees of freedom* will always relate to the number of axes found on a robot. The **axis** of a robot can be defined as the rotational section of the manipulator arm that rotates around the centerline of the body.

The **tooling** and **grippers** are not part of the robotic system itself; rather, they are attachments that fit on the end of the robot's arm. These attachments connected to the end of the robot's arm allow the robot to lift parts, spot-weld, paint, arc-weld, drill, deburr, and do a variety of tasks, depending on what is required of the robot.

The robotic system can also control the work cell of the operating robot. The **work cell** of the robot is the total environment in which the robot must perform its task. Included within this cell may be the controller, the robot manipulator, a work table, safety features, or a conveyor. All the equipment that is required in order for the robot to do its job is included in the work cell. In addition, signals from outside devices can communicate with the robot in order to tell the robot when it should assemble parts, pick up parts, or unload parts to a conveyor.

BASIC COMPONENTS

The robotic system has three basic components: the manipulator, the controller, and the power source. A fourth component, the end effector, is found on some robotic systems. These components are described in the following subsections.

Manipulator

The manipulator, which does the physical work of the robotic system, consists of two sections: the mechanical section and the attached appendage. The manipulator also has a base to which the appendages are attached. Figure 1–1 illustrates the connection of the base and the appendage of a robot.

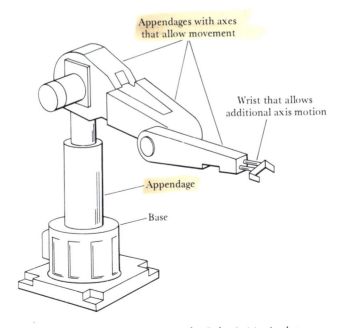

Figure 1–1 Basic Components of a Robot's Manipulator

The base of the manipulator is usually fixed to the floor of the work area. Sometimes, though, the base may be movable. In this case, the base is attached to either a rail or a track, allowing the manipulator to be moved from one location to another. For example, one robot could work at several presses, loading and unloading each press.

As mentioned previously, the appendage extends from the base of the robot. The appendage is the arm of the robot. It can be either a straight, movable arm or a jointed arm. The jointed arm is also known as an *articulated arm*.

The appendages of the robot manipulator give the manipulator its various axes of motion. These axes are attached to a fixed base, which, in turn, is secured to a mounting. This mounting ensures that the manipulator will remain in one location.

At the end of the arm, a wrist (see Figure 1–1) is connected. The wrist is made up of additional axes and a wrist flange. The *wrist flange* allows the robot user to connect different end-of-arm tooling (EOAT) to the wrist for different jobs.

The manipulator's axes allow it to perform work within a certain area. As mentioned earlier, this area is called the work cell of the robot, and its size corresponds to the size of the manipulator. Figure 1–2 illustrates the work cell of a typical assembly robot. As the robot's physical size increases, the size of the work cell must also increase.

The movement of the manipulator is controlled by actuators, or drive systems. The actuator, or drive system, allows the various axes to move within the work cell. The drive system can use electric, hydraulic, or pneumatic power. The energy developed by the drive system is converted to mechanical power by various mechanical

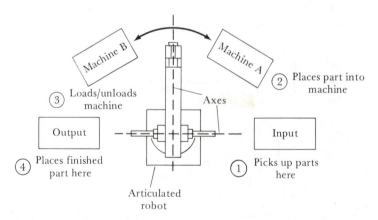

Figure 1–2   Elements of a Work Cell from the Top

drive systems. The drive systems are coupled through mechanical linkages. These linkages, in turn, drive the different axes of the robot. The mechanical linkages may be composed of chains, gears, and ball screws.

Controller

The controller, generally a microprocessor-based system, in the robotic system is the heart of the operation. The controller stores preprogrammed information for later recall, controls peripheral devices, and communicates with computers within the plant for constant updates in production.

Figure 1–3 illustrates a typical controller and manipulator that might be found in a robotic application. The controller is used to control the robot manipulator's movements as well as to control peripheral components within the work cell. Notice in the figure that the individual is programming the movements of the manipulator into the controller through the use of a hand-held teach pendant. This information is stored in the memory of the controller for later recall. The controller stores all program data for the manipulator movements. It can store several different programs, and any of these programs can be called up and run without reprogramming the robot.

The controller is also required to communicate with peripheral equipment within the work cell. For example, the controller has an

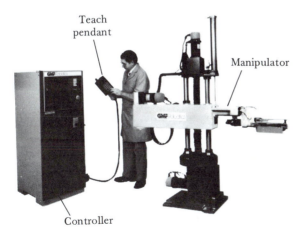

Figure 1–3 Programming Manipulator Motion into the Controller Using a Hand-Held Teach Pendant (Courtesy of GMFanuc Inc.)

input line that identifies when a machining operation is completed. When the machine cycle is completed, the input line turns on, telling the controller to position the manipulator to a preprogrammed location so that it can pick up the finished part. Then, a new part is picked up by the manipulator and placed into the machine. Next, the controller signals the machine to start operation.

The controller can be made from mechanically operated drums that step through a sequence of events. This type of controller operates with a very simple robotic system. The controllers found on the majority of robotic systems are more complex devices and represent state-of-the-art electronics. That is, they are microprocessor-operated. These microprocessors are either 8-bit, 16-bit, or 32-bit processors. This power allows the controller to be very flexible in its operation.

The controller can send electric signals over communication lines that allow it to talk with the various axes of the manipulator. This two-way communication between the robot manipulator and the controller maintains a constant update of the location and the operation of the system. The controller also controls any tooling placed on the end of the robot's wrist.

The controller also has the job of communicating with the different plant computers. The communication link establishes the robot as part of a computer-integrated manufacturing (CIM) system.

As the basic definition stated, the robot is a reprogrammable, multifunctional manipulator. Therefore, the controller must contain some type of memory storage. The microprocessor-based systems operate in conjunction with solid-state memory devices. These memory devices may be magnetic bubbles, random-access memory, floppy disks, or magnetic tape. Each memory storage device stores program information for later recall or for editing.

Power Supply

The **power supply** is the unit that supplies power to the controller and the manipulator. Two types of power are delivered to the robotic system. One type of power is the AC power for operation of the controller. The other type of power is used for driving the various axes of the manipulator. For example, if the robot manipulator is controlled by hydraulic or pneumatic drives, control signals are sent to these devices, causing motion of the robot.

For each robotic system, power is required to operate the manipulator. This power can be developed from either a hydraulic power source, a pneumatic power source, or an electric power source. These power sources are part of the total components used to position the manipulator.

Figure 1–4A illustrates the connection of a hydraulic supply to the base of the robot's manipulator. The hydraulic supply produces fluid under pressure. This fluid is sent to the various actuators on the manipulator. The fluid then causes the axis to rotate around the base of the robot.

Figure 1–4B illustrates a pneumatic supply, which places air under pressure. The pressurized air is connected to the manipulator, causing the axis to move in a linear fashion along the rail. The pneumatic supply is also connected to a drill and serves as the power for the rotation of the drill. Generally, the pneumatic supply is taken from the plant's supply, is regulated, and then is fed to the robotic manipulator's axes.

Figure 1–4C illustrates the electric supply. The electric motors can be either AC motors or DC motors. Electric pulse signals are

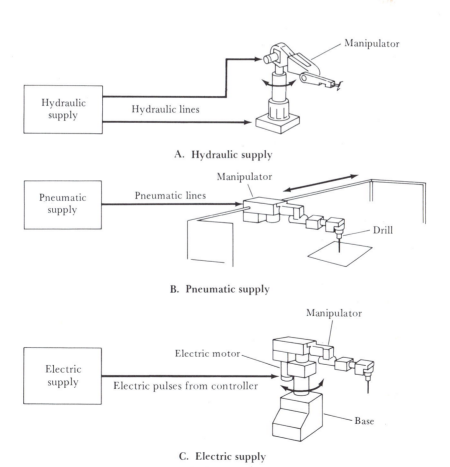

A. Hydraulic supply

B. Pneumatic supply

C. Electric supply

Figure 1–4 Types of Manipulator Power Supply

developed by the controller and sent to the electric motor on the manipulator. These electric pulses supply the motor with the necessary command information to cause the manipulator to rotate on the base of the robot.

In each of the three power systems for the manipulator's axes, a feedback-monitoring system is used. This system constantly feeds back positional data for each axis to the controller.

For each robot system, power is required to operate the controller as well as the manipulator's axes. This power can be developed from sources found in the manufacturing environment.

End Effector

The **end effector** or end-of-arm tooling (EOAT) found in most robot applications is a device connected to the wrist flange of the manipulator's arm. The end effector is used in many different situations in the production area; for example, it can be used for picking up parts, for welding, or for painting. The end effector gives the robotic system the flexibility necessary for operation of the robot.

The EOAT is generally designed to meet the needs of the robot's user. These devices can be manufactured by the robot manufacturer or by the owner of the robotic system.

The EOAT is the only component found on the robotic system that may be changed from one job to another. Figure 1–5A illustrates the robot connected to a water jet cutter, which is used to cut aside panels for auto production lines. The end effector can be changed as illustrated in Figure 1–5B, which shows the robot stacking parts onto a tray. In this simple process, the robot's end effector was changed, allowing the robot to be used in other applications. The changing of the end effector and the reprogramming of the robot allow this system to be very flexible.

ROBOT MOTION

Robot motion can be classified into two basic formats: the base travel and the axes control, which includes the arm motion, the wrist action, and the gripping action. These formats are described next.

Base Travel

The travel of the robot refers to the movement of the base. In most cases the base of the robot is permanently mounted to the factory floor. However, the travel of the base can also be regulated by mounting the base to rails or tracks, thereby making the robot

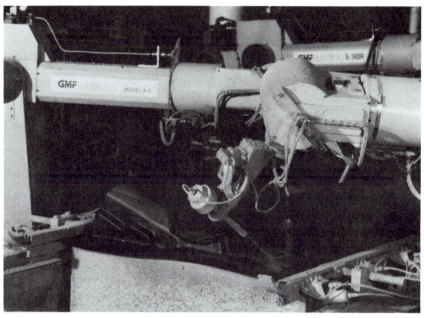

A. Water jet cutting

B. Stacking parts on a tray

Figure 1–5   Change of the End Effector So That a Robot Can Perform Different Tasks (Courtesy of GMFanuc Inc.)

movable. Figure 1–6A illustrates a mounting of the robot manipulator to a track.

For some jobs, the base may have to be mounted overhead or
at a slight angle. Figure 1–6B illustrates a typical overhead con-

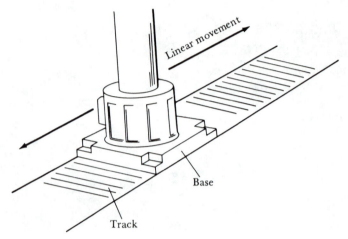

A. Track mount for linear travel

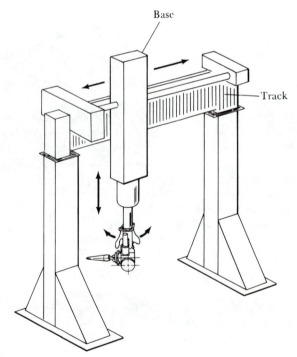

B. Overhead mount

Figure 1–6   Robot Manipulator Base Mounts

nection of the robot's base. This connection is called a gantry-mounted robot. The track that is used for the base allows the robot to reach positions that require work on the roof of a car. A spray-painting nozzle connected to the gantry-style robot would allow the robot to paint the roof of a car passing under the gantry.

Axes Control

As mentioned earlier, for some robots, the control of the robot's position is due to a constant feedback from the axes to the controller. This constant update is accomplished through devices known as servo systems, or simply servos. A **servo system** is a device that provides positional control over the axes of the robot. The servo contains command pulses that cause the motor to move, a feedback device that converts the positional data of the motor into electric pulses, and a feedback device that sends the positional information back to the robot controller. When the feedback pulses and the command pulses are the same, we say that the servo is *in position*.

With regard to axis control, robots used in the industrial environment are classified into two areas: non-servo-controlled robots and servo-controlled robots. Each type is discussed in the following subsections.

Non–Servo Control. Robots with **non–servo control** are identified as *bang-bang robots*. The movement of these robots' axes is stopped by a hard mechanical stop (hence the term *bang-bang*) placed in the travel path. Each time a new axis of travel is used, the robot's hard stops must be changed. These robots can have only three axes, or degrees of freedom: up/down, in/out, and left/right.

Figure 1–7 illustrates a non-servo-controlled robotic system. Notice in the figure that there is no feedback to the controller to identify the relative position of each axis. As the axis moves, it comes into contact with two detection switches S_1 and S_2. Detection switch S_1 indicates if the axis is at maximum travel, while S_2 detects maximum travel in the opposite direction.

Servo Control. In **servo-controlled robots**, the servo control allows the mechanics of the robot to communicate with the electronics of the controller. These robots also are classified as continuous path, servo-controlled robots and point-to-point, servo-controlled robots.

Figure 1–8 illustrates a servo-controlled robotic system. Notice that in this figure a feedback path is established between the servo motors and the controller. This feedback allows the controller to keep constant data stored for the location of each axis.

Continuous-path, servo-controlled robots allow a smooth path to be traced out by the axes of the robot. Generally, this path is a

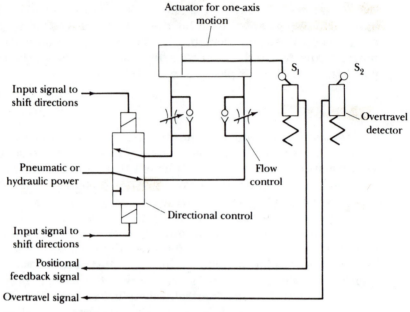

Figure 1–7 Typical Non-Servo-Controlled Robotic System

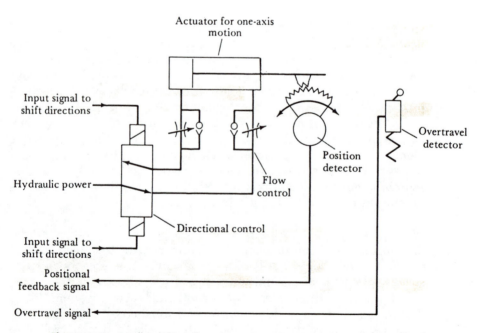

Figure 1–8 Feedback Developed in a Servo System to Identify the Position to the Controller

curved path. These robots require a great deal of memory storage within the controller. The curved path developed by these robots works well in the area of welding, spray painting, and assembly.

The *point-to-point, servo-controlled robot* allows movement of the axes between a starting point and a preprogrammed ending point. In this robot, all axes move at the same time in order to reach the programmed point. The movement of this robot in its path of travel is very jerky. Applications for the point-to-point robot are found in machine loading and unloading.

Each of these styles of axes movement can be found in different robots. And in many cases, the point-to-point movement and the continuous-path movement are found in the same robotic package.

ROBOT TECHNOLOGY LEVELS

Industrial robots have three levels of technology: low technology, medium technology, and high technology. Each of these different classes of robotic operation has unique characteristics. The different classes relate to different needs within the manufacturing area. Each class is discussed in the following subsections.

Low-Technology Robots

Low-technology robots are used in the industrial environment for jobs like machine loading and unloading, material handling, press operation, and very simple assembly operations. Each of these different jobs requires that special features be built into the robot's operating characteristics. These characteristics are discussed in the paragraphs that follow.

Axes. The first characteristic of the low-technology robot is the number of axes found on the system. Low-technology robots generally have between two and four axes of movement. They are basically non-servo-controlled robots. Thus, mechanical stops are used at the end of each axis of travel. The axes motions are generally up/down. Also, a reach axis is usually found on these robots. Finally, these robots have the capability of rotating a gripper. Figure 1–9 shows a simple low-technology robot with the features just described.

Payload. One of the important concerns of the low-technology robot is the maximum payload found at the end effector. **Payload** is the load capacity (weight) a manipulator can position. This weight is measured at the center of the wrist flange of the robot, as indi-

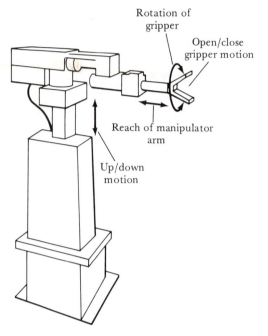

Figure 1–9   Non-Servo-Controlled, Low-Technology Robot

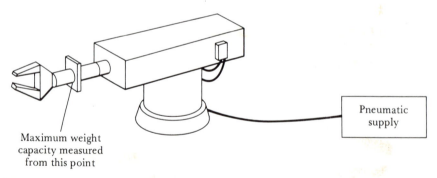

Figure 1–10   Measurement of the Weight Capacity of a Robot

cated in Figure 1–10. In low-technology robots, the maximum amount of weight can range from 3 to 13.6 kilograms.

Cycle Time.   The time it takes the robot to move from one location to the next is also very important in any robot operation. This time is generally called the **cycle time**. The cycle time of the robot depends on two factors: the payload and the length the manipulator's arm must travel. Low-technology robots will generally have very high cycle times. The movement of the axes to various positions can take from 5 to 10 seconds. Movements of this type could increase the production time required to fabricate a part.

Accuracy.   Another important characteristic of any robotic system is the accuracy of the system. The accuracy of the robotic operation describes how closely a robot can position its payload to a given programmed point. Related to the robot's accuracy is its **repeatability,** which describes how often the robot doing the same program can repeat its payload to a given point.

These two features, accuracy and repeatability, are very important in any robotic operating system. Many of the low-technology robots are capable of developing high accuracy and repeatability. The accuracy of the low-technology robots ranges from 0.050 to 0.025 millimeter.

Actuation.   The methods of driving the robot's axes, called the *actuation,* differ among various low-technology robots. The actuation can be accomplished through the use of pneumatics, hydraulics, or electricity. The major reason for using one system or the other is the direct cost. Because of the lower cost of electric motors, some of the low-technology systems are now using electric stepper motors.

Pneumatics gives the robot the ability to hold medium-weight payloads. Hydraulic systems allow the manipulator to hold and position heavier payloads. The electric drive is the easiest of the three to control.

Controllers.   The low-technology robots are identified as hard automation systems. **Hard automation systems** are controlled by the setting of various cams or sequencing valves. Each time a new job must be done, a hard automation system must have its cams and sequencing valves reset.

With low-technology robots, each new job must be re-preprogrammed. Low-technology robots have a very limited amount of memory space available for program information. Generally, only timing information and sequencing information can be programmed.

Controllers for low-technology robots can be either electronically controlled or air-logic-controlled. Once the new microprocessing power, memory, and servo control now available are applied to these robots, their flexibility can be increased.

Medium-Technology Robots

Medium-technology robots are used primarily for picking and placing and for machine loading and unloading. These robots are a bit more sophisticated than the low-technology robots, as the characteristics described next indicate.

Axes.   In the majority of cases, the medium-technology robots have a greater number of axes on the manipulator than the low-technology robots have. These robots also have a larger work cell, which means that the axis travel is greater.

The number of axes on these robots is increased because of the addition of a wrist. The wrist connected to the end of the robot's arm gives the system two or three additional axes. The wrist also gives greater maneuvering power to the manipulator. The wrist allows the robot to develop movements that the low-technology robot is not capable of. With the addition of the wrist, these robots may have five or six axes of motion.

The basic medium-technology robot has three axes, or degrees of freedom, as shown in Figure 1–11A. With the addition of a wrist, as illustrated in Figure 1–11B, three additional axes are added to the robot. These axes are called the *roll axis,* the *bend axis,* and the *yaw (rotational) axis* of the wrist.

With these additional axes, the medium-technology robot is an excellent choice for servicing two machines in the loading and unloading of parts. Figure 1–12 illustrates a medium-technology robotic system servicing two machines in such an operation.

Payload.   Medium-technology robots have a greater payload capacity than the low-technology robots. These robots are capable of handling weights at the end of the wrist ranging from 68 to 150 kilograms. With this increased payload, these robots are able to

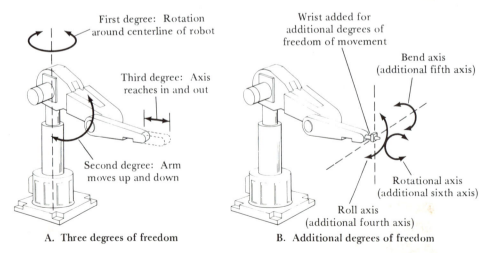

A. Three degrees of freedom   B. Additional degrees of freedom

Figure 1–11   Axes for a Medium-Technology Robot

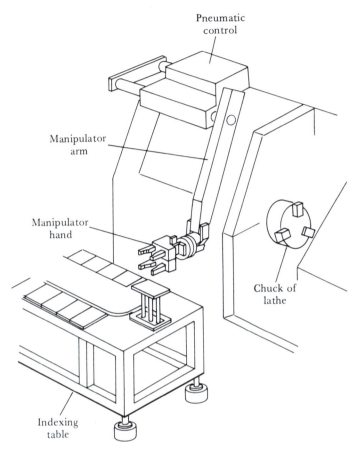

Pneumatic
control

Manipulator
arm

Manipulator
hand

Chuck of
lathe

Indexing
table

Figure 1–12 Loading/Unloading Operations of a Medium-Technology Robot (Adapted with permission from GMFanuc Inc.)

replace the worker in situations where a constant lifting of heavy parts is required.

Cycle Time. For a medium-technology robot, a movement in the reach axis of 25 to 65 centimeters takes 1.0 second to execute. In a rotation around the center axis, the robot can move at a rate of 150 centimeters per second. However, these robots will have greater cycle times than the low-technology robots. The complex jobs the medium-technology robots are able to perform and the payload at the wrist flange are responsible for the increased cycle time of the robot's operation.

Accuracy. Because of the increased number of axes on medium-technology robots, their accuracy and repeatability are not as good as the accuracy and repeatability of low-technology robots. The main reason is that the several axes must converge on a point. The error that is developed because of several axes having to be positioned creates additional error in the accuracy of the robot. Remember that the low-technology robots only move one axis at a time to reach a programmed position. However, the medium-technology robots are capable of repeating their positional data to meet the requirements of the job.

The accuracy of the medium-technology robots can range from as low as 0.2 millimeter to as high as 1.3 millimeters. These ranges are typical for various medium-technology robots.

Actuation. Medium-technology robots are driven by two types of motors: hydraulic or electric. These two types of actuation are used because of the heavy payloads.

Controllers. The controllers found on the medium-technology systems are microprocessor-based. That is, the system can have either an 8-bit or a 16-bit microprocessor. With the increased power of the processor, memory size is also increased. Thus, positional data can be programmed into these robots, stored in memory, and recalled for later jobs.

Several jobs can be stored in the controller's memory, thus allowing the robot to service two or more machines in the loading and unloading of parts. These robot controllers also allow the user to change or edit program information.

High-Technology Robots

High-technology robots can be used for material handling, press transferring, painting, sealing, spot welding, arc welding, and a variety of different tasks found in the manufacturing environment. These robots represent the state of the art in robotic systems. Their characteristics are described in the following subsections.

Axes. High-technology robots have from 6 to 9 axes. As the technology increases, the number of axes may increase to 16 or more. These robots' movements are meant to resemble human movements as closely as possible.

Figure 1–13 illustrates a robotic paint system that incorporates the use of seven axes. These axes are controlled by hydraulic operation because of the safety concerns arising from the robot's working around paint fumes.

Payload. The payload for these robots remains about the same as the payload for medium-technology robots, from 68 to 150 kilograms.

Figure 1–13   Seven-Axes, High-Technology Robot Used for Spray Painting (Courtesy of GMFanuc Inc.)

Cycle Time.   The cycle times of the high-technology robots are about the same as the cycle times of the medium-technology robots. The main difference is that these robots have a cycle time for each axis. Thus, the cycle times for these robots are based on a composite time of all the axes moving together.

Accuracy.   The accuracy of high-technology robots is greater than the accuracy of the other two classes of robots. The increase is due to the positional data that is fed back to the controller, thus allowing the servo system to develop accurate positional data. The specifications of the accuracy of many of the high-technology robots are from 1 to 0.4 millimeter. This range gives the robot the necessary accuracy to do many of the jobs required in the industrial environment. With the increase of accuracy, the repeatability of the robot also increases.

Actuation.   The main drive systems found on high-technology robots are electric drive motors and hydraulic drives. There has been a trend in the industry to use electric drive systems. The electric drives are easy to position, but with these drives, the payload capabilities are reduced.

Controllers.   The real achievement in high-technology robots is the controller. These controllers are powered by 16-bit microprocessors. The memory capacity for these units is on the order of 1M bytes (M represents millions).

The controller contains positional control circuits that are themselves small microprocessors. As stated earlier, these positional circuits maintain the robot in the preprogrammed position.

The major feature of the high-technology robot that distinguishes it from the other classes of robots is the ability of the controller to interface with additional peripheral devices in the robot's environment. These devices, such as sensors, vision systems, and off-line programming, are all part of the increased capabilities of high-technology robots.

SUMMARY

The robot is made up of three basic components: the manipulator, the controller, and the power source. End effectors are found on complex units as well. The manipulator comes in a variety of shapes and sizes; it does the physical work of the system. The controller controls the manipulator, its position, and its movements. The robotic system's controller also has the ability to interface with peripheral devices in the robot's environment, such as sensors, vision systems, or master computers. The power source provides power to the manipulator and to the controller.

End effectors are components that are attached to the end of the robot's manipulator arm. The end effector is used to pick and place parts, arc-weld, spot-weld, or spray-paint. These devices allow the robot to interface with the manufacturing environment.

The robot's degrees of freedom describe the number of axis movements. Depending on the technological level of the robot, the number of axes can range from 2 to 16.

Robot motion includes base travel and axis control. With regard to axis control, robots may be non-servo-controlled or servo-controlled. The latter units are more flexible, have more axes, and provide continuous-path or point-to-point control.

The robotics industry has developed three basic classifications of robots: low-technology robots, medium-technology robots, and high-technology robots. Each class of robots has the ability to perform certain jobs in the workplace. Also, each class is defined by specific characteristics of axes, payload, cycle time, accuracy, actuation, and controllers.

≣ KEY TERMS

accuracy	manipulator
actuator	medium-technology robot
appendage	non-servo-controlled robot
axes	payload
controller	power supply
cycle time	preprogrammed location
degrees of freedom	repeatability
drive systems	robot
end-of-arm tooling (EOAT)	servo-controlled robot
end effector	servo system
gripper	tooling
hard automation system	travel of robot
high-technology robot	work cell
low-technology robot	wrist

≣ QUESTIONS

1. What is another name for the axes found on the manipulator?

2. What is the maximum number of axes that may be found on a manipulator?

3. At what point on the manipulator does a gripper attach?

4. Name the four basic components of a robotic system.

5. How can the manipulator be made portable? Draw a sketch to support your answer.

6. Identify the name of the area in which a manipulator can move without obstruction.

7. Which component on the manipulator develops axis motion?

8. What component on the robotic system is the heart of the system?

9. What two components within the robotic system does the controller monitor?

10. What system can interface with the robotic system to aid in the manufacturing process?

11. Identify the three basic power supplies found in robotic systems.

12. What is the name of the style of robotic system that is mounted in an overhead manner?

13. What component in the robotic system is used to control the axis position?

14. What is the major difference between a servo-controlled robotic system and a non-servo-controlled robotic system?

15. What is the range of the number of axes that may be found on low-technology robotic systems?

16. From what point on the manipulator is the maximum amount of payload measured? Draw an example to support your answer.

17. What is the term used in robotics to describe the manipulator returning to the programmed location?

18. What type of actuation is found on many of the low-technology robot systems?

19. What are the two types of controls that are used in low-technology robotic systems?

20. What is the major component that is found on medium- and high-technology robotic systems that is not found on low-technology robotic systems?

21. What is the range of payload that can be handled by the medium-technology robotic systems?

22. What electronic advantage do the medium- and high-technology controllers have over the low-technology controllers?

23. What type of actuation is found on the high-technology controllers?

24. Name several of the various types of peripheral devices that can be interfaced with high-technology controllers.

2

Basic Features of the Manipulator

OBJECTIVES

Upon completing this chapter, you should be familiar with:

— Manipulator arm geometry,
— Wrist rotation,
— Manipulator drive systems,
— Work envelopes,
— Manipulator mounting.

INTRODUCTION

The robotic system has been defined as having three basic components: the manipulator, the controller, and the power source. Each of these components must operate in order for the robot to perform work. This chapter describes the basic features and operation of the manipulator. Succeeding chapters present details of the controller and the power source.

The robot manipulator is classified according to its arm movement. And the arm movement is described by four coordinate systems: Cartesian coordinates, polar coordinates, cylindrical coordinates, and articulate coordinates. Each system is detailed in this chapter.

Other topics covered in this chapter include wrist motion, manipulator drive systems, the work envelope of the robot, and manipulator mounting. Additional axes of motion can be given to the robot by placing a wrist at the end of the manipulator. The wrist can add an additional three axes to the robot. These axes provide the roll motion, the yaw motion, and the pitch motion of the wrist.

The motion of the manipulator is driven by either electric, hydraulic, or pneumatic power sources. These power sources are converted through actuators to provide the power for driving the motion of the robot manipulator.

The range of axis motion defines the work envelope of the robot manipulator. The work envelope is the area in which the axis motion can operate without any obstruction.

The final topic discussed in this chapter is the mounting of the robot manipulator. The manipulator can be mounted from an overhead gantry, in an angle position, or in an upright position.

COORDINATE SYSTEMS

In many of the high-technology manipulators, coordinate systems are used to identify the motion of the robot. These coordinate systems are identified by three axes: the X axis, the Y axis, and the Z axis. Figure 2–1A illustrates the three coordinate planes. Each vector identifies a plane of reference. These planes are often called *frame of reference* to describe the manipulator's motion. The central starting location for the coordinate system is identified as the origin of the coordinate system. It must be noted that the origin of the coordinate system can move in any direction, thus generating a new frame of reference for the coordinate system. Notice in Figure 2–1B that the origin of the coordinate system has shifted, thus causing new direction for the X axis, the Y axis, and the Z axis. The direction of the three axes always depends upon the user's point of view.

Many high-technology controllers employ coordinate systems. The coordinate systems that are used are identified as the world coordinate system, the tool coordinate system, and the user coordinate system.

World Coordinate System

The world coordinate or world frame system is identified as a fixed location within the manipulator. Notice in Figure 2–2 the location of the world coordinate system. Here, the fixed location is identified in the center of the manipulator's shoulder. All points within the manipulator's range of motion or location within the world can be referenced from this point.

The X axis of travel moves the manipulator in an in-and-out motion, while Y axis motion causes the manipulator to move side to side. The Z axis causes the manipulator to move in an up-and-down motion.

Tool Coordinate System

The tool coordinate or tool frame system can be used by a robot to identify its location within the work envelope. The tool coordinate

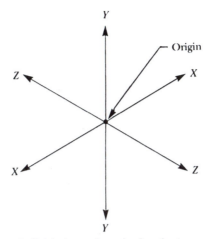

A. Origin (central starting location)

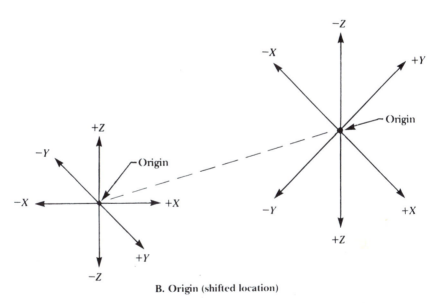

B. Origin (shifted location)

Figure 2–1   Coordinate System Identifying X, Y, and Z Axes

system is referenced from the center of the tool being mounted on the faceplate of the manipulator. Notice in Figure 2–3A that the origin of the coordinate system is located on the faceplate of the manipulator. From this location, we can see the direction of the X, Y, and Z axes. All movement of the manipulator will be referenced from this location in space. In Figure 2–3B, a tool has been added to the faceplate of the manipulator. Notice that this addition of the tool has moved the origin of the coordinate system to a new location.

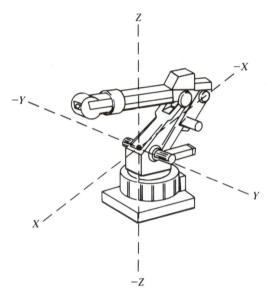

Figure 2–2 World Coordinates Identified on an Articulate Style Manipulator

This new location will be the new point of reference for the movement of the manipulator.

User Coordinate System

User coordinates or user frames are identified as the coordinate system located on the part to be handled by the manipulator. Notice in Figure 2–4A that the user's part has a coordinate system identified in relationship to the X, Y, and Z axes of the part. The identification of these axes is in relationship to the origin of this coordinate system.

The user coordinate system for the manipulator is used to align the coordinate system located on the faceplate of the manipulator with the coordinate system of the user's part. Once these two systems are aligned, the manipulator will then move along the user's part coordinate system motion. Figure 2–4B illustrates the alignment of the two systems. It must be remembered that the manipulator will move along the same coordinate reference as the user's part coordinate system. Also, it should be pointed out that, if a tool is added to the faceplate of the manipulator, the tool coordinate system and the user coordinate system must be utilized in order to move the manipulator correctly.

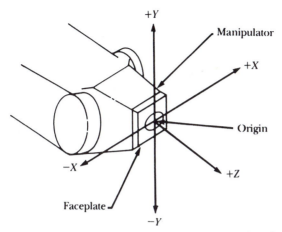

A. Coordinate system located at center of faceplate of manipulator

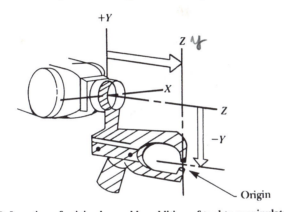

B. Location of origin changed by addition of tool to manipulator

Figure 2–3   Tool Offset Causing Origin of Coordinate System To Move

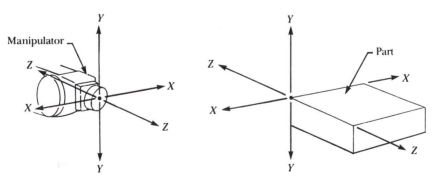

A. Coordinate system located at   **B. Manipulator's coordinate system**
center of faceplate of manipulator   **aligned with part's coordinate system**

Figure 2–4   User Coordinate System Aligning Manipulator Coordinates with Part's
             Coordinates

MANIPULATOR ARM GEOMETRY

Robots are classified according to the type of axis movements needed to complete a task. The four classes of **arm geometry** within the *coordinate systems* are Cartesian, cylindrical, polar, and articulate. In each of these geometries, the arm moves through space, causing certain axis movements. These movements are identified as the arm geometry. Each type of arm geometry is described in the following subsections.

Cartesian Coordinates

From a mathematical standpoint, the **Cartesian coordinate system** identifies three basic axes, or three planes. These planes are the X plane, the Y plane, and the Z plane. Each of these planes relates to movement of the arm axis from a point of origin. Generally, this point of origin in the robot is the **centerline** of the robot.

Figure 2–5 illustrates the three axes of the Cartesian coordinate robot. Notice that at the convergence of the three axes is the center point, or the point of **origin**.

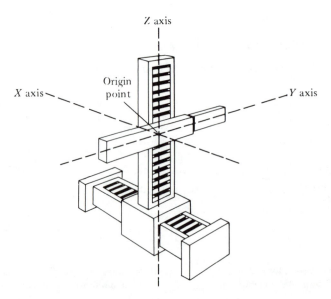

Figure 2–5 Cartesian Coordinates X, Y, and Z

The X, Y, and Z planes identify the motion of the manipulator. The manipulator can move through space only in the X, Y, and Z planes to reach its target. Figure 2–6 illustrates the manipulator's arm movement in the Cartesian coordinate robot. In the figure, the arm moves along the Z *axis* in a linear, up-and-down motion from point A to point B. When the arm is moving along the Y *axis*, it is moving in a linear, in-and-out (reaching) movement from point B to point C. This linear motion is defined in terms of the center point of the robot. The arm can also move along the X *axis* (not shown in Figure 2–6). This movement is a linear, side-to-side motion.

A robot designed through the use of the Cartesian coordinate system can move only within these three X, Y, and Z planes. Any positional point between the limits of these axes is impossible to reach with this robot. Generally, the Cartesian robot operates within a rectangular pattern. Hence, the work envelope of this robot is rectangular.

Movement between two of the axes in the Cartesian coordinate system is impossible. For example, with reference to the manipulator shown in Figure 2–6, a direct movement from point A to point C is impossible with a Cartesian coordinate robot. The robot's movements can only be in the X, Y, and Z directions.

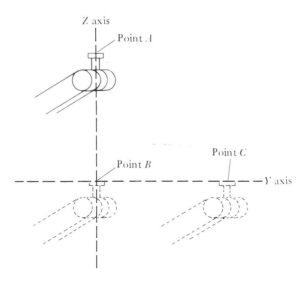

Figure 2–6   Manipulator Arm Moving from Points A to B to C in the Cartesian Coordinate Robot

A robot with Cartesian coordinate axes is basically used for point-to-point applications. So, the Cartesian coordinate geometry is a useful system for robots that are loading and unloading material into machines. The Cartesian coordinate robot is restricted to low-technology robotic systems.

Cylindrical Coordinates

The cylindrical coordinate system incorporates three degrees of freedom, or three axes: the *theta* (θ, Greek letter theta), or *rotational, axis;* the Z, or *up-and-down, axis;* and the R (reach), or *in-and-out, axis*. This robotic system is called the cylindrical system because the work envelope of the robot traces out a cylinder.

Figure 2–7 shows a cylindrical coordinate robot. The theta axis defines the rotation around the base. In many robotic systems,

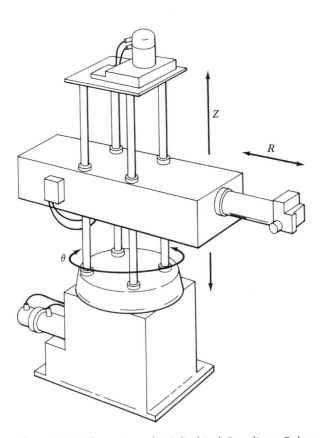

Figure 2–7   Three Axes of a Cylindrical Coordinate Robot

this rotational base has about 300° of movement. The remaining 60° is identified as a **dead zone,** which is a safety zone of operation.

The reach, or R, axis allows the robot to reach for various pieces of work. The reach of the robot can range from 500 to 1500 millimeters, depending on the robot and the task.

The Z axis allows the robot to move in an up-and-down direction. The travel limit for this axis can be from 100 to 1100 millimeters, depending on the task of the robot application.

Polar Coordinates

The **polar,** or spherical, **coordinate system** also has three axes: the theta (θ) axis, the reach (R) axis, and the beta (β, Greek letter beta) axis. This robotic system is called the spherical system because the robot's manipulator traces out a sphere as its work envelope.

Figure 2–8 illustrates the movement and the axes of the polar coordinate robot. The polar robot's *theta* and R *axes* basically define the same movements as the theta and R axes of the cylindrical system. The *beta axis* allows the robot's entire arm to bend in an up-and-down direction.

Articulate Coordinates

The **articulate,** or jointed, **coordinate system** is identified by three axes: the theta (θ) axis, the W (upper arm) axis, and the U (elbow) axis.

Figure 2–9 illustrates the movement and the axes of the articulate robot. Notice that the robot has two axes that have the

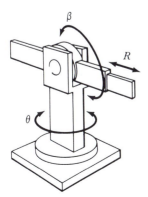

Figure 2–8   Three Axes of a Polar Coordinate Robot

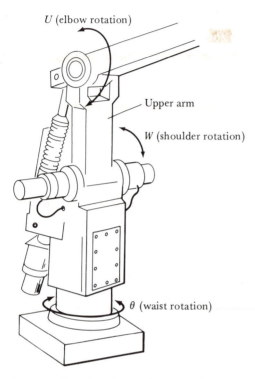

Figure 2–9   Three Axes of an Articulate Coordinate Robot

capability of bending. This action gives the articulate robot greater flexibility in the workplace.

The articulate robot of Figure 2–9 has rotation around the base. This rotation is provided by the *theta,* or *waist, axis*. The bar above the theta axis is the upper arm of the robot. Its axis is the *W*, or *shoulder, axis*. This *W* rotational axis provides movement similar to the movement allowed by the beta axis in the polar coordinate robot. The *U*, or *forearm, axis* allows a bending movement similar to the *W* axis bending. This axis is also called the *elbow axis*.

The articulate robot is one of the most popular robots for industrial application. The main reason for its popularity is the flexibility of its arms.

THE SCARA MANIPULATOR

In many of the new applications that are used in mechanical and electronic assembly, a new type of manipulator is employed. This

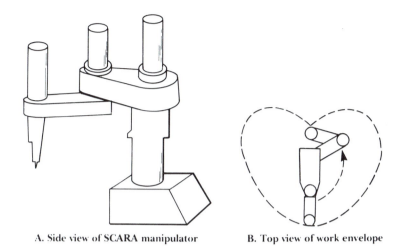

A. Side view of SCARA manipulator   B. Top view of work envelope

Figure 2–10   SCARA-Style Manipulator

manipulator is identified as a horizontal-arm selective compliance assembly robot arm, or SCARA, manipulator. Figure 2–10A shows the side view of a SCARA manipulator, while Figure 2–10B illustrates the work envelope of a SCARA manipulator.

Notice that the SCARA manipulator has at least three axes of movement: the theta axis, or the waist rotational axis; the upper arm axis, a Z axis used for up-and-down motion; and, finally, the wrist rotational axis. Typically this type of mechanical manipulator's fast axes motion will allow it to carry only very light payloads weighing 2.5 kilograms to a maximum of 25 kilograms.

As stated, these manipulators are very fast, but can only carry minimal payloads in operation. A typical SCARA-style manipulator has a theta work envelope of 300° and can travel that distance in one second.

WRIST ROTATION

The four different coordinate systems describe the planes in which the manipulator is able to maneuver. Each of these systems gives only limited axis movement. The addition of a wrist at the end of the robot's arm extends the mobility of the robotic system. The addition of the wrist also increases the dimensions of the work envelope.

Figure 2–11 illustrates the movement and the axes of a wrist. With the majority of wrists in use today, an additional two or three axes are added to the robot's mobility. Figure 2–11 shows a wrist

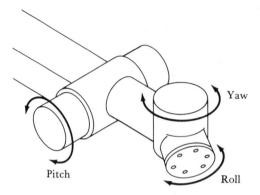

Figure 2–11 Wrist Axis Motions

that develops three additional axes of movement: the yaw axis, the pitch axis, and the roll axis.

The **yaw axis** describes the wrist's angular movement from the left side to the right side. This motion can range from a 90° movement to a 270° movement, depending on the design of the wrist.

The **pitch axis** describes the wrist's rotational movement up and down. The angular motion of the pitch can range from merely a few degrees of motion to 270°, depending on the application of the wrist.

The **roll axis** describes the rotation around the end of the wrist. The roll axis can provide rotation up to 360°. With an end effector connected to the roll axis, a full 360° of rotation can be achieved.

The addition of these extra axes allows the robotic system to be very flexible. The **degrees of rotation** that the wrist provides are variable. For example, the yaw can have 270° of travel, the pitch can allow 90° to 110° of travel, and the roll, as stated earlier, can add 360° of rotation. Or a wrist can develop two rolls and no pitch. But however the wrist axes are designed, the addition of the wrist to the robot's arm allows the end effector to reach into areas that could not be reached by robots using only one of the four coordinate systems for the arm. The flexibility of the system is thus increased with the different wrist designs used in the robotic operation.

MANIPULATOR DRIVE SYSTEMS

The manipulator must have some type of drive system that will cause the mechanics of the robot to operate. In industrial robotic systems, there are three basic drive systems: hydraulic, pneumatic, and electric. Each of these systems will be discussed in detail in later chapters. This section gives a brief introduction to their operation.

In about 15% of the robotic systems found in the industrial environment, pneumatic power is used for the axes' driving power. Pressurized air is supplied through lines to cylinders, causing the air pressure to be transformed into mechanical axis movement.

The pneumatic system is the least expensive of the three drive systems. The major problem with the pneumatic system is the limited amount of pneumatic power that can be converted into lifting power, or *torque*. A typical pneumatic system is only capable of lifting 3 to 4.5 kilograms.

The pneumatic-operated manipulator is generally found in applications of simple assembly, die-casting operations, material handling, and machine-loading or -unloading operations. The applications for this manipulator are limited because of the payload restrictions. A comparison of the payload for robots with pneumatic, hydraulic, and electric drive systems follows:

	Low-technology robot	Medium-technology robot	High-technology robot
Payload	3–4.5 kg (6.61–10 lb)	22.7–56.7 kg (50–125 lb)	3–80 kg (6.61–176 lb)
Drive system	pneumatic	hydraulic	electric/ hydraulic

Hydraulic drive systems are used in about 20% of the industrial robots on the market today. The hydraulic system supplies fluid under pressure to a cylinder. The pressurized fluid entering the cylinder causes the cylinder to extend or retract, depending on the direction of the fluid in the cylinder. The cylinder is connected to the manipulator's axis. The motion of the cylinder then causes the manipulator's axis to move.

The typical hydraulic-operated manipulator is capable of lifting payloads in the area of 22.7 to 56.7 kilograms. The hydraulic robot manipulator is used in applications such as material handling, press loading and unloading, spot welding, arc welding, die casting, and spray painting.

The hydraulic system used in robotics has two disadvantages. First, the hydraulic system is one of the most costly drive systems. Second, in use of the hydraulic-operated system, the hydraulic fluid may leak onto the shop floor. This leakage could cause unsafe working conditions around the work envelope of the manipulator.

The torque delivered by the hydraulic system is the best of the three systems. Thus heavy lifting of parts is best done by hydraulic-drive robots.

The electric drive systems use either AC or DC electric motors. These motors are connected to the manipulator's axes through a gear reduction process. The gear reduction process allows the electric motor to develop the torque necessary for the robot to lift heavy payloads.

The typical payloads for electric drive systems range from 3 to 80 kilograms. The electric drive systems are very versatile in operation. The electric motors allow smooth start-up of payloads and smooth deceleration and stopping of the robot at its final position.

Typical applications for electric manipulator drive systems are in assembly, arc welding, spot welding, machine loading and unloading, material handling, and deburring. Moreover, they have a host of other uses within the plant. Also, newer and safer motors are being used in spray-painting applications. These motors do not generate sparks, and thus, they eliminate the worry of an explosive situation.

WORK ENVELOPES

The most important characteristic to any individual working near the robotic system is how far the robot can reach. The reach of the robot is defined as the **work envelope** of the mechanical manipulator. All programmed points within the reach of the robot are part of the work envelope.

As described previously, the Cartesian coordinate robot has a rectangular envelope. The work envelope of a cylindrical coordinate robot is cylindrical. The work envelope of the polar coordinate robot is spherical. And the articulate coordinate robot has a tear-shaped work envelope.

Figure 2–12A illustrates the side view of the work envelope of an articulate coordinate robot. Notice that the work envelope of the robot reaches to the floor and reaches above the horizontal plane of the top of the robot's forearm.

Figure 2–12B illustrates the top view of the work envelope for the same articulate coordinate robot. Notice that the robot's envelope extends in a semicircle around the robot. A dead zone is located at the rear of the manipulator's path.

From Figure 2–12B we see that the work envelope of this articulate robot has a range of motion of 285°. The starting point of the robot's motion is from the zero reference point. The manipulator can then move 142.5° in the positive direction and move 285° to the

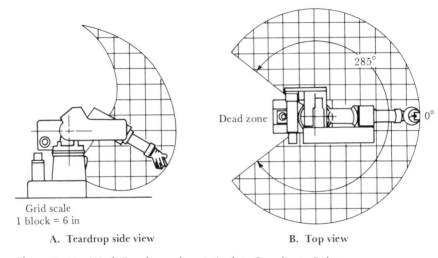

A. Teardrop side view   B. Top view

Grid scale
1 block = 6 in

Figure 2–12   Work Envelope of an Articulate Coordinate Robot

other end of the work envelope. Notice that the full extension of the robot can reach 41 inches.

Figure 2–13 illustrates a typical work envelope for a cylindrical coordinate robot manipulator. In Figure 2–13A, the manipulator is profiled from a side view. The shaded area illustrates the maximum reach of the arm of the manipulator. Notice that the robot can only move 550 millimeters in the Z plane. Also, the manipulator is already 790 millimeters above the floor of the workshop.

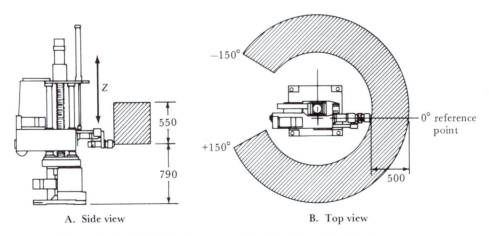

A. Side view   B. Top view

Figure 2–13   Work Envelope of a Cylindrical Coordinate Robot

Figure 2–13B illustrates the top view of the work envelope. This view shows the movement of the manipulator around the base. As the figure shows, the manipulator can move from a zero reference point at the center, and it can move in a clockwise direction to the +150° mark. The manipulator can also move from the +150° point to the −150° point. If the arm of the manipulator is fully extended, the work envelope is increased by 500 millimeters. This extension of the work envelope must be taken into account when the user is planning for the robot's installation. Note that these views, side and top, allow the user to identify all the reach capacities of the robot during its operation.

For the description (or drawing) of a work envelope, the measurements are taken from the wrist flange. Thus if any end-of-arm tooling is attached to the robot, the work envelope size will increase by the length of the end-of-arm tooling. For instance, suppose the work envelope of the articulate coordinate robot shown in Figure 2–12A is 192.02 centimeters. Then, when a spot-welding gun is added to the wrist flange, the work envelope's arc will increase by approximately 47.24 centimeters. This increased dimension is very important to the people working around the robot. For safety of the workers, this increased dimension must be considered in the design of the robot's installation.

MANIPULATOR MOUNTING

The manipulator is the movable arm of the robot. As defined earlier, the manipulator consists of a base and an appendage. Manipulator mounting refers to the mounting of the base.

The manipulator can be mounted in a variety of different locations and positions. For instance, the manipulator can be mounted in an upright position, or it can be hung from a gantry.

If the robot is to be stationary, then the base can be secured to the floor of the plant. This mounting can be accomplished in several fashions. Figure 2–14 illustrates a typical stationary-mounting method for the manipulator. The baseplate is secured by floor anchoring cement to the floor of the plant. The baseplate is then attached to the leveling bolts, and the manipulator's base is attached to the baseplate.

A method of moving the robot to different workstations is illustrated in Figure 2–15. Here, the robot is mounted on tracks. When the user wants to move the robot from one job location to the next, the user simply pushes the robot manipulator down the track. When the robot is in position, pins are placed into the floor to keep

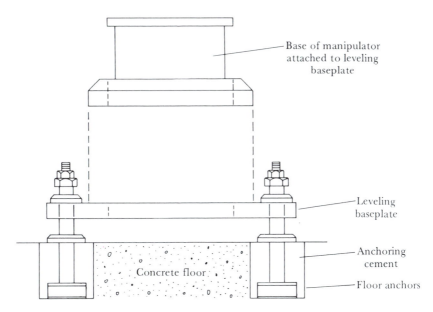

Figure 2–14 Mounting of a Robot Base to a Floor

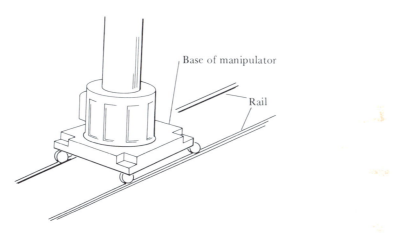

Figure 2–15 Manipulator Base Attached to a Rail for the Movement of the Manipulator from One Job to Another

the robot from moving any farther along the track. This positioning is very important when repeatability is needed in production runs.

A manipulator may also be mounted to a wall at angles. This mounting procedure is used when space is a problem or when the

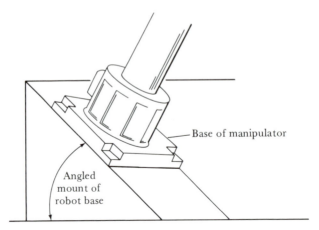

Figure 2–16   Articulate Robot Mounted at an Angle for an Extension of the Work
Envelope

angle required for performing a task cannot be reached by a manipulator mounted to the floor.

Figure 2–16 illustrates the mounting of a manipulator at a slight angle. This mounting allows the reach of the manipulator to be extended. For example, if the manipulator must reach into deep pockets in an arc or spot welding application, a wall mounting at a slight angle gives the manipulator the reach it needs.

Each of these three mountings provides a different work envelope for the robot. And different work envelopes will require different safety measures and different positioning of the workpieces. Thus, the mounting of the manipulator has the following three major aspects.

The first concern is for the personnel. No person working around the manipulator should come into contact with the manipulator when it is in operation. Thus, a clear and well-defined work envelope should be identified to ensure the safety of plant personnel.

The second aspect concerns the positioning of the robot manipulator with respect to the controller. The manipulator and the controller should be as close to each other as possible. Increasing the distance between the controller and the axis motors will develop line losses through the cables. Possible miscommunication between the controller and the manipulator's drive mechanisms may result. In the hydraulic system, the distance between the manipulator and the hydraulic pumps could create problems in the fluid pressure delivered to the axis drive mechanism.

The third aspect involves the alignment of the robot and the worktable. If the robot is to do arc welding, for example, and the robot is not anchored down, the robot can creep during normal operation. This slight creeping can cause the robot manipulator to move out of position from the workpiece, thereby resulting in incorrect welds. So, for this type of work, a well-secured, stationary mount is required.

SUMMARY

The robotic system can have four different arm geometries. These geometries allow different axis movements. The four basic geometries are the Cartesian coordinate system; the cylindrical coordinate system; the polar, or spherical, coordinate system; and the articulate, or jointed, coordinate system. Each coordinate system defines a unique work envelope for the robot.

The Cartesian coordinate robotic system has three basic axes of movement: the X axis for side-to-side motion, the Z axis for up-and-down motion, and the Y axis for in-and-out reach. Its work envelope is rectangular. The cylindrical coordinate robot is designed to develop a work envelope shaped like a cylinder. The manipulator can move in rotation around the base, which is a movement in the theta axis. It can move in an up-and-down motion, which is a movement in the Z axis. And it can reach in and out, which is called a movement in the R axis.

The polar coordinate manipulator also has a theta rotation and an R axis movement. But this manipulator has the ability to swing the R axis in a top-to-bottom radius, which is a movement in the beta axis. So, the work envelope of the polar coordinate robot is spherical. The articulate coordinate robot provides additional bending movements for the axes so that the manipulator can better reach various locations within the work envelope. For the articulate robot, the theta axis provides rotation, or bending, around the waist. The W axis provides rotation around the shoulder. And the U axis allows bending around the forearm. The work envelope of the articulate robot is shaped like a teardrop.

Additional axis motion can be obtained by the addition of a wrist to the end of the robot's arm. Wrists can provide up to three additional movements for the robot's manipulator. These movements are the pitch, the yaw, and the roll of the wrist.

Three types of drives are currently available for the robot manipulator: pneumatic drives, hydraulic drives, and electric drives. The type of drive system used for the robot manipulator depends on the speed and the lifting power required by the task.

The work envelope of any robotic system is important in describing the complete operation of the robot. Work envelopes differ according to the manipulator's movements. The work envelope of the cylindrical coordinate robot has a cylindrical shape. The work envelope of the polar coordinate robot is spherical. And the work envelope of the articulate coordinate robot is tear-shaped.

The mounting of the manipulator depends on the task required of the robotic system. The robot can be mounted in an upright position; it can be mounted from a gantry; or it can be mounted at an angle. These different mounting positions relate to the task the robot must perform.

KEY TERMS

arm geometry
articulate coordinate system
beta axis
Cartesian coordinate system
centerline
cylindrical coordinate system
dead zone
degrees of rotation
frame
origin
pitch axis
polar coordinate system
roll axis
R, or in-and-out, axis

SCARA manipulator
theta, or rotational, axis
theta, or waist, axis
tool coordinate system
user coordinate system
U, or forearm, axis
work envelope
world coordinate system
W, or shoulder, axis
X axis
yaw axis
Y axis
Z axis
Z, or up-and-down, axis

QUESTIONS

1. Name the four coordinate systems that are used with manipulators.

2. Name the three axes that are found on a Cartesian coordinate manipulator. Draw a sketch to support your answer.

3. State the type of axis motion that a Cartesian coordinate manipulator has.

4. Name several applications in which a Cartesian coordinate manipulator can be used.

5. In what technology group is the Cartesian coordinate manipulator found?

6. What axis on the cylindrical coordinate manipulator is the rotational axis?

7. How many degrees of freedom are there for a cylindrical coordinate manipulator?

8. What is the typical size of the safety zone for cylindrical coordinate manipulators?

9. What is the typical reach for cylindrical coordinate manipulators?

10. What axis on a polar coordinate manipulator develops the bending motion?

11. What type of work envelope does the polar coordinate manipulator trace?

12. In what coordinate system does the manipulator have an elbow axis, a forearm axis, and a shoulder rotation?

13. How many additional degrees of freedom can be given to a manipulator by adding a wrist?

14. Draw the yaw motion for a robot's wrist, and identify the degrees of motion.

15. What is the typical payload range for electric manipulators?

16. Define the term *work envelope.*

17. Draw the side view of a typical work envelope for the articulate manipulator.

18. What is the effect on the work envelope when an end effector is added to the manipulator?

19. Identify three types of mountings that the manipulator can have.

20. What drive systems are generally found on low-technology manipulators?

21. What is the typical payload range for medium-technology manipulators?

22. What drive system is used in many of the spray-painting manipulators?

23. Draw the location of the origin point of the Cartesian coordinate manipulator.

24. Define the differences between world coordinates, user coordinates, and tool coordinates.

3

Major Internal Components of Controllers

OBJECTIVES

Upon completing this chapter, you should be familiar with:

— General features of controllers,
— Characteristics of controllers by level,
— Input power supply board,
— Master control board,
— Memory boards,
— Servo control board,
— Interfacing board,
— Signal path in the controller.

INTRODUCTION

The controller of the robotic system houses the electronic circuits for manipulator control. The controller contains the robot's microprocessor, the various printed circuit boards needed for the robot's operation, and the control circuits for the motors.

The controller of the robotic system can range from a very low level of technology, requiring only cam and switch setting for each program, to a very high level, containing state-of-the-art microprocessors. In the majority of robotic systems made today, the manufacturer uses state-of-the-art microprocessor technology in the controller.

Low-technology controllers only develop sequencing of the various cams. These controllers have no memory capacity. Medium-technology controllers have limited memory capacity and low-power microprocessors. These controllers are not easily reprogrammed. Controllers for high-technology robotic systems have powerful microprocessor systems and large memory capacity for later reprogramming. High-technology controllers also have the capacity for controlling up to nine axes of the manipulator. Furthermore, they provide the user with the ability to edit program information and change jobs by calling up new program numbers. The high-technology controller is the main focus of this chapter.

In many of the medium- and high-technology controllers, the internal components include several printed circuit boards. These boards contain all the necessary electronic circuits to support the capabilities of the controller. The boards generally found in the controller are the input power supply board, the master control circuit board, the memory board, the servo control circuit board, and the interfacing circuit board. Each of these printed circuit boards serves an important function in the total operation of the robot and the controller. Each board will be briefly described in this chapter; more detailed discussions of the boards are given in later chapters.

GENERAL FEATURES OF CONTROLLERS

Figure 3–1 illustrates the schematic for a typical servo-controlled robot using a hydraulic system. Notice that the block diagram for the control system contains two components: program memory and a comparator, which allows the comparison of programmed data with feedback from the moving axes.

Outside the controller is the teach pendant. The **teach pendant** is used during the teaching of programmed points. The **cathode ray tube (CRT)** is used to display positional data of the program so that the operator can identify the location of the end-of-arm tooling as it is jogged to various locations within the work envelope.

During operation, an output is developed from the controller to a servo valve. This electronic signal controls the on-and-off switch of the servo valve. The servo valve directionalizes the flow of fluid into or out of the hydraulic cylinder. Notice also in Figure 3–1 that hydraulic fluid is used for the operation of the cylinder.

A feedback device allows the controller to identify the position of the stroke of the manipulator arm. This data is constantly fed back to the controller for constant updating of the positional data. Positional correction is made by the controller and fed to the servo valve. Thus all data related to the operation of the robot is processed through the controller.

As in any type of control system, cables are used to connect the controller to the servo valves. And piping is used from the servo valves to control the cylinder (the hydraulic actuator). Thus, the distance between the controller and the physical unit is very important so that cable and fluid leakages are controlled.

Since the controller is operating in environments where the temperature could go above the limits of the electronic circuitry, cooling devices are required in the controller system. Heat ducts

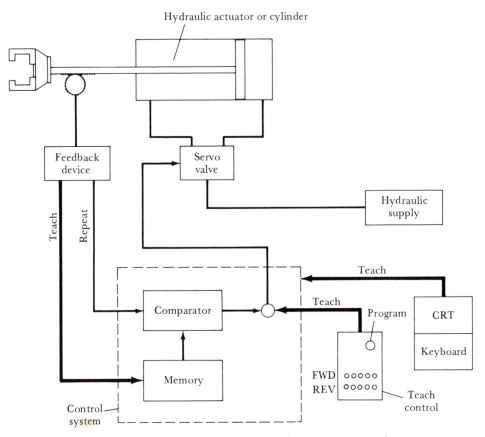

Figure 3–1 Schematic for a Servo-Controlled, Point-to-Point Robot

are also used; they contain fans that provide circulation around the controller's internal components.

Figures 3–2A and 3–2B illustrate one type of cooling process. Notice in Figure 3–2B that a heat transfer duct is used to circulate the air. Air is drawn in through the bottom of the unit by fans at the top of the duct. Fans draw the cooling air through the duct. Heat from the internal components is transferred to the duct. The hot air of the internal cabinet is exchanged with cooling air from the duct. When the environment around the controller does not provide cooling air for the heat exchange, air conditioning units can be provided for the controller cabinet.

The duct system of the controller provides a closed environment that is contamination-free. The internal controller components are thus protected from any contamination in the plant's environment.

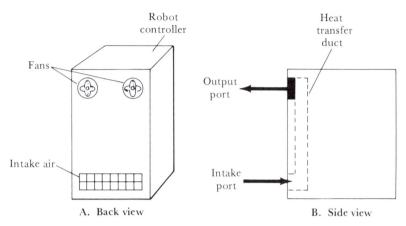

Figure 3–2   Cooling Devices for a Controller

CHARACTERISTICS OF CONTROLLERS BY LEVEL OF TECHNOLOGY

The controller is a major component of the robotic work cell. The controller has the responsibility for controlling the manipulator's movements as well as controlling peripheral devices within the work cell.

Three types of robot controllers are used in the manufacturing environment. These types are classified by the level of technology. Thus, there are low-technology controllers, medium-technology controllers, and high-technology controllers.

The low-technology controller is used on systems where the job of the robot remains constant and where the robot has limited steps to perform in the operation. These controllers are not easily reprogrammed to perform another operation. The medium-technology controller has only a limited memory capacity for storing the robot's program information. It also has a low-powered, 8-bit microprocessor to perform the various single-axis motions. The high-technology controller has a large memory capacity and a high-powered microprocessor, 16 or 32 bits. Also, it is able to control the motion of several axes at one time. The following subsections describe the characteristics of each level of controller.

Low-Technology Controllers

Low-technology controllers may be reprogrammed, although their reprogramming is a difficult task. Remember that the ability to be reprogrammed is one of the important characteristics of the robotic work cell. Low-technology controllers control movement through cam-timing signals. So, the controller can be reprogrammed for another task through the movement of the various cams to different locations. Generally, the plant programmer has the responsibility of reprogramming the low-technology controller. This process is accomplished by physically moving the cams and the hard stops that are located in the manipulator's path of travel. This reprogramming process is a long and tedious task that can mean many lost hours of production. So, in practice, low-technology controllers usually are not reprogrammed.

Low-technology controllers have two disadvantages. First, they do not have any internal memory for storing the point location of the robot's manipulator movement. Second, they do not have any microprocessor-based system in which electronic command signals can be used for axis movement control.

Medium-Technology Controllers

Medium-technology controllers are generally found with robotic systems that have two to four axes of movement for the manipulator. These controllers have a microprocessor-based system and some type of memory capacity. Instructions from the microprocessor are sent to the various axes to develop controlled movement. These controllers also have limited memory capacity so that programs can be stored for later recall.

Medium-technology controllers also have a limited input/output signal capacity for communicating with peripheral devices around the controller. For example, consider a robot being used for a machine-loading/unloading operation. When the chuck (the attachment for holding a part) of the machine tool is ready for a part, it sends a signal to the robot controller to place a part into the chuck. When the machining cycle is complete, the machine tool sends an output signal to the robot controller to come into position and pick up the part from the chuck. Figure 3–3 illustrates this communication between the controller and the machine tool.

Figure 3–3   Input/Output Signals between a Controller and a Machine Chuck

Through its microprocessor, the medium-technology controller can develop servo control over the manipulator's axis movement. The servo control will continue to send feedback pulses to the processor for comparison with the command movement pulses stored in controller memory. If the command pulses and the stored memory pulses are the same, the robot manipulator is said to be *in position*. In this case, the robot will stop movement and execute its informational data.

Whenever a command is received by the microprocessor, a time interval elapses while the command is executed. This time interval will delay each movement of the robot. This delayed action is a common trait of medium-technology robots. That is, they are slow to react and can only support movement of one axis at a time, because the controller requires microprocessor time to execute each axis movement command.

Figure 3–4 is a block diagram of the operation of medium-technology robotic systems. The steps in this operation are as follows:

1. A command to move an axis is initiated.
2. The microprocessor receives command signals from memory.
3. The signal is sent to the motor control circuits.
4. The signal is sent to the motor.

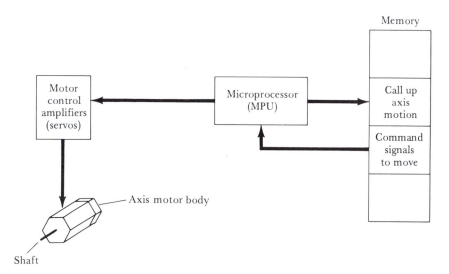

Figure 3–4 Operation of a Medium-Technology Robotic System

In this operation, the microprocessor requires time to fetch the data from memory, to read that data, and, finally, to execute the data. The time it takes the microprocessor to process these commands slows the robot's movements.

The medium-technology controller can be reprogrammed once a job has been finished. Unlike the low-technology controller, the medium-technology controller stores its programmed information in solid-state memory. The one limiting factor of the medium-technology controller is the amount of memory space available for program information. The limited amount of memory may only be enough to store one or two operational programs at a time. Thus, after completion of one program, the operator can call another program from memory. But if additional programs are required, the controller must be reprogrammed.

High-Technology Controllers

High-technology controllers generally have large memory capacities, a microprocessor and a co-microprocessor, servo control for manipulator movements, and up to 32 input/output signals for communication of the controller to peripheral devices.

Figures 3–5A and 3–5B illustrate block diagrams of the internal components of a typical high-technology controller. The controller is divided into five major sections: power supply, interface, axis drive boards, option boards, and microprocessor. The power supply block (Figure 3–5B) has the task of converting the plant's AC voltage into internal DC operating voltage for the controller. The interface block (Figure 3–5A) is responsible for establishing communication links between the end effector on the manipulator, the peripheral devices located in the work cell, and various external controller components. The axis drive boards are responsible for the operation of the axis drive systems. The axis drive boards are generally servo control circuits that supply motion command signals to the axis drives. The option block is used to supply additional memory space for the controller and to supply separate interfacing ability to the controller. For example, the block labeled "vision interfacing board" is responsible for interfacing the signal from the vision system to the microprocessor section. The main block of the controller is the microprocessor. This block is the main computing section of the controller. All data required for the operation of the robotic work cell are commanded from this section.

High-technology controllers also have the ability to be reprogrammed within a very short period of time. This reprogramming ability allows the operator to call up different programs when different parts are presented to the robot's manipulator. For example,

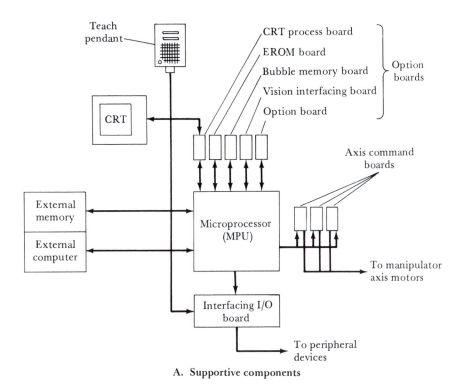

A. Supportive components

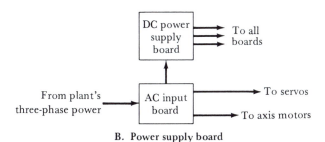

B. Power supply board

Figure 3–5   Block Diagram of Some Internal Components in a High-Technology Controller

consider a six-axis robot used for spot-welding applications on an assembly line. Each different car body coming down the line has a separate style (code) number that must be read by the controller. And each code number calls up a different program from the controller for the execution of the spot-welding path. All of the different welding programs are contained in the controller's memory.

High-technology controllers can manipulate up to nine axes at a time. Thus, the controller can command all the axes at one time so that they move simultaneously. Therefore, the manipulator arrives at a location more quickly than a medium-technology manipulator, and the movement of the axes is a smooth path to the target location.

In many high-technology controllers, two microprocessing units (MPUs) are used to call stored data from the memory location and to execute the data. These two microprocessor units, the MPU and the co-MPU, are shown in Figure 3–6. The *microprocessor* can call data either from the memory location or from the various input and output lines, and it can execute all such data. The microprocessor-based systems generally have high levels of microprocessor families that allow for short data execution time. The *co-microprocessor* (co-MPU), the second MPU found in many systems, is used to calculate arithmetic functions of the controller, such as the positional data of the manipulator for the servo operations. The basic function of the co-microprocessor is to take some of the work from the main microprocessor, allowing the main MPU to do work important to the operation of the controller and the manipulator.

Another important concept of the controller is **multi-tasking**. Multi-tasking capabilities allow the controller to process two tasks at the same time. This means the processor could be calculating positional data, while at the same time reading the I/O of the controller. This important characteristic allows the robot controller to move toward being the controller for the entire work cell.

The memory of high-technology controllers has a very large capacity for storage of data. Like any computer memory, the controller's memory is made up of combinations of *bubble memory, random-access memory* (RAM), and *read-only memory* (ROM). These memory devices are solid-state components that have the capacity for storing all the positional data and the operating control programs.

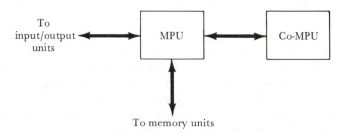

Figure 3–6 MPU and Co-MPU Used to Control Robot Operation

The controller's memory can range from 32K to 1M. bytes of storage capacity. This memory space is very valuable. So, high-technology controllers have the ability to download their memory to peripheral devices, where information is stored for later use. The peripheral devices include floppy disks, magnetic tapes, and bubble memory cassette adapters, as shown in Figure 3–7. Information taken from the controller and stored on these devices opens up additional memory space in the controller for storing new programs. Another purpose of this downloading operation is to keep a permanent record of the program being executed. If, for some reason, the controller were to lose its memory, a simple upload from the external memory device is all that is needed to put the controller back into action.

Many of the new high-technology robotic systems are tied into a totally automated manufacturing system. This system uses the **computer-aided design/computer-aided manufacturing (CAD/CAM) procedure** to interface with the controller. The flow of this procedure is illustrated by the block diagram of Figure 3–8. Programs for the robot's operation are generated on a CAD terminal. This process, called **off-line programming**, allows all of the robot's movements to be programmed via a computer terminal. All of the robot's positional data are also programmed from this terminal. Once these programs

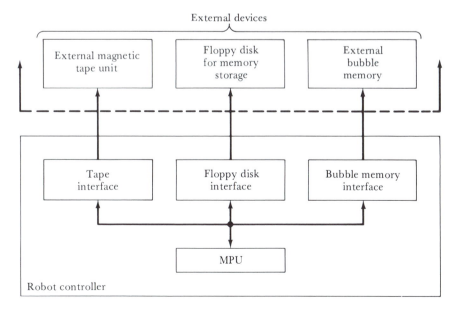

Figure 3–7 Peripheral Memory and an MPU for a High-Technology Controller

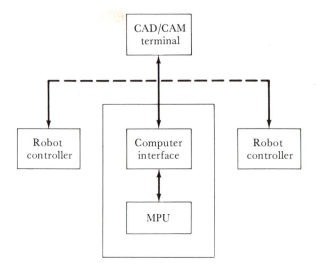

Figure 3–8 CAD/CAM System Interfacing through the Computer Interface Port

are generated, they are downloaded to the robot's controller memory. The data is transferred over a computer link, called a **data highway,** between the controller and the CAD terminal. The program information generated on the CAD terminal can now be touched up by an operator on the manufacturing floor.

High-technology controllers also have the ability to interface with *sensing devices,* such as vision and tactile sensors, which expand the robot's capabilities. Different sensors allow the robot to distinguish different parts in a bin, identify bad components, or inspect a hole for its depth. The information developed from the sensor allows the controller to accurately position the end-of-arm tooling. Additional information on sensing devices will be presented in Chapter 12.

INPUT POWER SUPPLY BOARD

The **input power supply board** consists of two major sections: the three-phase input AC voltage from the plant and the internal DC power supply, which is used for operation of the controller's other internal circuits. Figure 3–9 illustrates these two circuit boards and their connections. Each board is described in the following subsections.

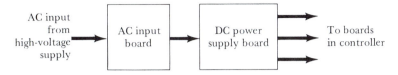

Figure 3–9   Block Diagram of Input Power Supply

Input AC Voltage

The AC voltage is supplied from the three-phase voltage, usually 480 volts, 3 phase, AC, operating within the plant. This high AC voltage must be stepped down to a usable level for the controller. The reduced AC voltage levels are in the range of 100 to 200 volts AC, depending on the operation of the internal control circuits. The reduced voltage is then converted to lower DC voltage levels and to lower AC voltage levels for the operation of the motors on the manipulator or of the actuators on the hydraulic manipulator.

Figure 3–10 illustrates the relationship between the two boards. As shown in Figure 3–10A, the AC input power is developed at the main breaker. The AC supply voltage is then fed through fuses F_1, F_2, and F_3, which protect the controller against overload conditions. Also at the input side of the AC voltage are surge protection circuits, which are responsible for protecting the controller's circuitry in case of high-voltage surges on the input power line.

The AC voltage is passed through relay contactors to the servo system. The AC voltage is also supplied to a step-down transformer. Here, the AC voltage is reduced to a usable level for the operation of the robotic work cell. The resulting voltage levels are 100 volts AC for the general operating circuits and 24 volts AC for the general DC power supply circuits.

Internal DC Power Supply

The output of the secondaries of the step-down transformers is passed to DC rectifiers, which are located on the DC power supply board, as shown in Figure 3–10B. The rectifier converts the 24 volts AC into a DC voltage. The next block of the power supply is the filter circuit. This circuit converts the pulsating DC voltage from the rectifier into smooth DC operating voltage. The regulator circuits are responsible for maintaining a constant output voltage to the load. That is, if the input voltage increases or decreases within a tolerance range, the regulator circuits keep the output voltage at a constant level. The next block in the figure is the voltage divider

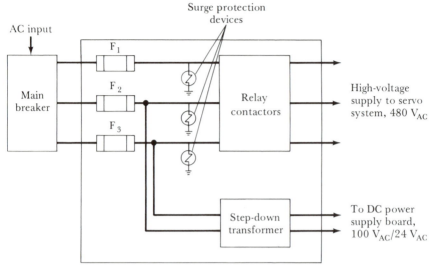

A. Distribution of AC input power

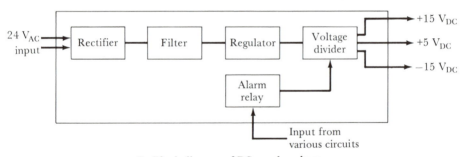

B. Block diagram of DC supply voltage

Figure 3–10   Input Power Supply for a High-Technology Controller

circuit. This circuit converts the 24 volts DC output voltage into the various levels needed by the controller's components.

The final block shown in Figure 3–10B is the alarm circuitry. This circuit is placed in the power supply as a safety circuit. Before DC power can be applied to the various internal circuits of the controller, the alarm circuitry checks to see whether there are any electronic problems in these circuits. If there are no problems, then the DC power is supplied to the controller. If there is a problem, then the DC power is shut down, and the controller is not allowed to start up.

Generally, the internal DC voltage levels are from +5 to +15 volts DC. Each of the DC lines has circuit fusing in case overloads are developed in the circuits. Overloads can be created from the overheating of a motor, a short in a solid-state device, or problems in the AC input circuitry. Each of these defects could cause increased current flow through the power supply.

The tolerances of the voltages of the DC power supply board are critical to the overall operation of the controller. If these voltages drop or rise above certain tolerance levels, the controller's operation can suffer. Thus, in many high-technology controllers, regulators are used with the outputs of the DC power supplies. The regulators keep the DC output voltage at a constant level.

Also found in many of the high-technology controllers are circuits called alarm relays, as noted earlier. These alarm circuits monitor the operation of the controller's circuits. For example, if low or high voltages are generated in the controller's circuitry, the alarm relay that monitors the circuits causes the entire controller to shut down.

Since many of the controllers contain large quantities of RAM, memory batteries are being placed in the controllers. These batteries are used as backup power sources to retain the memory in RAM in case of power failure within the controller.

MASTER CONTROL BOARD

The **master control board** is the main control board for the controller. Mounted on this board are all of the control electronics necessary for the operation of the manipulator, the peripheral devices, and the input/output signal control. The control of the electronics is handled by the MPU of the controller. The MPU is therefore also located on the master control board. Included with the MPU is all of the memory required for normal operation of the controller.

Figure 3–11 illustrates a typical master control board and the support components that might be located on it. Notice that two memory boards are connected to the master control board. These memory boards are the **executive memory board** (executive ROM, or EROM), which contains all of the operating instructions for the controller, and the **bubble memory**, which stores the user's operating program for the robot.

Also working in conjunction with the master control board are support printed circuit boards. One of these boards is used for control of a cathode ray tube (CRT) display, where program information and operating conditions of the robot can be displayed. The sensor interface printed circuit board is used to interface outside sensor devices to the controller. For example, a vision system for the robot would be connected through this printed circuit board. This board,

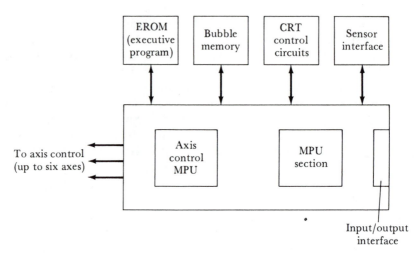

Figure 3–11 Master Control Board

then, is the major communication link between the master control board and the sensor components.

Since the master control board must communicate with other printed circuit boards in the controller, an input/output communication port is provided. This port is responsible for sending command signals from the master control board to internal controller components.

The following subsections discuss the major component on the master control board—the MPU—and the lines, called buses, that carry information to and from the MPU. The memory board, servo control board, and interfacing board are described in later sections of this chapter.

Microprocessing Unit

The main component on the master control board is the **microprocessing unit (MPU)**. The MPU controls the total operation of the robotic system. Figure 3–12 illustrates the basic components of the MPU system: the executive memory circuits, two input/output (I/O) circuits, the read/write memory for storage of program information, and the **central processing unit (CPU)**, which includes the program counter (PC), the arithmetic logic unit (ALU), and the internal register (IR). Each of these components is described in the following paragraphs.

The executive memory circuits hold the control programs. The **control program** is connected directly to the MPU, and it identifies for the MPU the different routines that the MPU must perform.

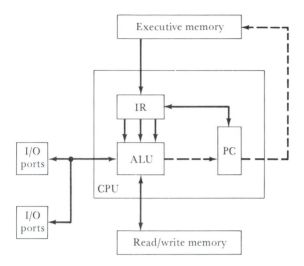

Figure 3-12 Basic Components of the Microprocessor

For example, the control program tells the MPU when to branch to subroutines, when to jump to other program statements, or when to add or subtract numbers in the ALU. The **arithmetic logic unit (ALU)** performs named arithmetic operations, as commanded by the executive program.

Also located in the CPU is the **program counter (PC)**. The PC keeps track of which step of the program's instructions the MPU is executing. When branches or jumps occur in the program, the PC is set forward or backward so that it keeps track of program execution at all times.

The **internal register (IR)** of the microprocessor is used to store data as commanded by the ALU. The IR receives information from the executive memory and stores this data until the ALU calls for it.

The **read/write (R/W) memory** stores the program path that makes up the robot's movements. Also stored in this memory are the general parameters for total system operation of controller and program. The separate memory units in the read/write memory are generally referred to as **registers**. The registers in the controller's memory circuits, for example, keep track of the parts count, the point when the controller should cease an operation, and the signal parameters used for the I/O signal operation.

The MPU must communicate with additional electronic circuits of the controller. These circuits are the **input/output (I/O) circuits**. The I/O circuits send the information that has been processed by the MPU to circuits in the controller.

Buses

The data information carried by the I/O circuits is sent over lines in the circuitry called **buses**. Figure 3–13 illustrates the connection of the buses used in the microprocessor system. Two buses are associated with the MPU: the address bus and the data bus.

In order for the MPU to communicate with internal electronic circuitry on the master control board, it must address various locations. The **address bus** is the communication link between the MPU and the desired address location. For example, suppose the MPU wants to address memory location 3FFF. This address is sent out on the address bus to the specified memory location. The MPU can then communicate to this memory location that it will either read or write to this location. The addressing process is the only method by which the MPU knows what section it should communicate with. The second bus associated with the MPU, the **data bus,** is used to transfer data to various circuits on the master control board.

Additional circuits outside the MPU are required for storing this data until the proper cycle time of the MPU is reached. These outside peripheral devices are called **latches**. Figure 3–13 illustrates two types of latches used with the MPU: the address latch and the input/output (I/O) data latch. The **address latch** receives an address location from the MPU and temporarily stores this data until the MPU commands the information to be sent. The command signal is processed through a line called the *strobe signal*. The **I/O data latch**, like the address latch, is another storage device used by the MPU for temporary storage of data. The I/O latch stores data being

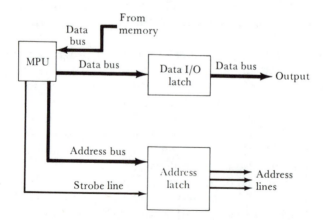

Figure 3–13   Address and Data Buses

sent by the MPU to peripheral circuits or stores data being sent from the peripheral circuits to the MPU.

Figure 3–14 illustrates a typical flow process for a situation when the MPU wishes to fetch data from another circuit in the controller. In the figure, the MPU is requesting data from a memory location. The MPU first addresses the memory location through the address latch. This data is placed on the address bus and held in the address latch until the MPU commands the address latch to address the desired memory location. Notice in the figure that the address location is 05. The address latch then sends the address to the address decoder in the memory circuitry. The address decoder selects the correct memory location—in this case, address 05.

A control signal is also processed from the MPU to the address location. This command signal tells the memory circuit whether data will be read from memory or written into memory by the MPU. In this example, the MPU wants to read data from memory. At the READ command, the data from memory location 05 is placed on the data bus and sent to the I/O data latch. At the commanded time

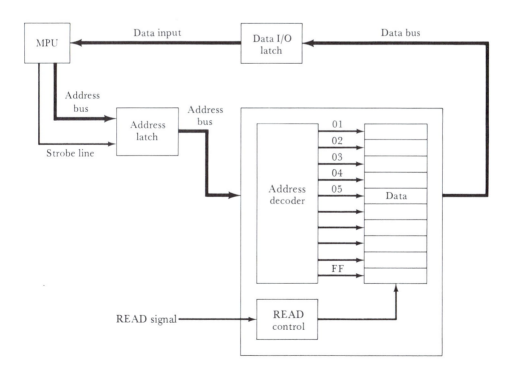

Figure 3–14 Transfer of Data to the Data Bus

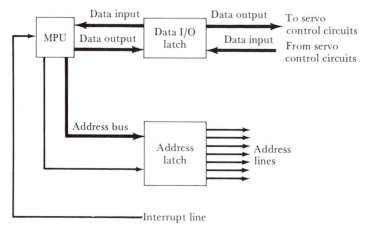

Figure 3–15   Input Data from the Servo Control Circuits

from the MPU, the data from memory location 05 is transferred to the MPU.

This fundamental process is carried out by the MPU to read and write data to locations in the robot controller. In the same manner, the MPU can address any location in the controller or in peripheral devices to command the operation of that component.

Figure 3–15 illustrates the opposite action of the MPU: information being sent from the servo units to the MPU. Data for this operation is waiting on the input data bus of the MPU. Before this data is transferred to the MPU, an interrupt signal is generated. The **interrupt signal** tells the MPU that data is waiting to be processed. Once the interrupt signal is received by the MPU, it will finish its arithmetic operations and call the waiting data from the servo units. The information is then processed by the MPU and returned to the servo units.

The microprocessor in many controllers has 8 data lines for communication with the microprocessor. With the development of new technology, the microprocessor can have up to 16 or 32 data lines for information that is to be processed. The increased size of the data buses increases the power of the processing system.

MEMORY BOARDS

One of the major differences among the three levels of controllers is in their ability to be reprogrammed. The reprogramming information is stored in the medium- and high-technology controllers in random-access memory (RAM), read-only memory (ROM), and bubble memory. These **memory boards** store the data necessary for the operation of the controller and for the programmed paths of the manipulator.

The data stored in memory are not words like the words we are used to reading but, rather, are digital codes. **Digital codes** break down various commands into bits of digital information. The term **bit** is short for "binary bit." The binary bit is a code of 1s or 0s. These codes identify whether the condition is on or off.

Memory space is conserved and the MPU operates faster when the stored information is arranged into patterns of 8 bits. This pattern is called a **byte** of digital information. Thus, bytes of program information are stored in the controller's memory. These bytes are then placed on the MPU data bus and sent to various locations in the controller.

Figure 3–16 illustrates a typical **memory map** for the controller. Each of the memory locations has a special purpose. For example, 7000 (7K) bytes of information are stored in the *programmable read-only memory* (PROM) for the control programs or the executive programs. The information in this location is always saved. If the power in the controller is removed, the contents of the main memory will remain.

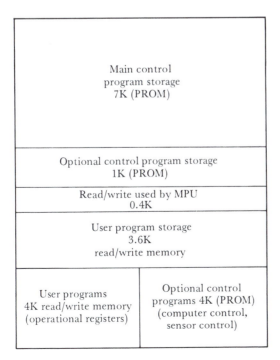

Main control program storage 7K (PROM)
Optional control program storage 1K (PROM)
Read/write used by MPU 0.4K
User program storage 3.6K read/write memory

User programs 4K read/write memory (operational registers)	Optional control programs 4K (PROM) (computer control, sensor control)

Figure 3–16 Controller's Memory Map

About 1K bytes of PROM memory are reserved for the MPU's memory. This memory location stores arithmetic operations when an overflow exists in the program's operation. A read/write memory location of about 0.4K is also provided for the MPU.

Additional memory space is reserved for positional data. **Positional data** is the information required for manipulator movement. When the robot is placed into the mode of operation called repeat, the controller calls up the various positional points from memory and executes these points in the same order in which they were taught. That is, the addressing process calls up this information in a sequential order. If this memory space becomes full, the programmer must download some information to external memory devices such as tapes, bubble cassettes, and floppy disks. The program can be uploaded to memory when it must be run again.

The external memory devices keep records of the controller's programs. This storage space gives the controller added flexibility. Different jobs can be stored and later put in operation by simply uploading program information.

Parameters for the controller's operation are stored in the memory referred to as **operational registers**. These registers contain the system's operating conditions. For example, suppose the robot is loading and unloading parts from a machine tool. After 100 parts have been worked on, the robot is to stop. So, there must be some memory location in which the programmer can keep track of the count of the number of parts worked on. This count is kept in the 4K of user program memory.

Servo gains, input and output peripheral signal control, and special offset data for picking up parts can all be stored in this 4K area of user program memory. When this information is needed by the operating program of the controller, it can be called up and executed by the microprocessor.

The memory for the controller is called **nonvolatile memory**. That is, if the AC power for the controller is removed, the data in the nonvolatile memory will remain. In order for the memory to remain, the controller must supply power to the memory circuits. The power for these circuits can come from batteries in the controller or from a small voltage developed from the AC input and provided to the DC power supply.

SERVO CONTROL BOARD

Servo control boards are found in medium- and high-technology controllers. These boards are used to control the positioning of the

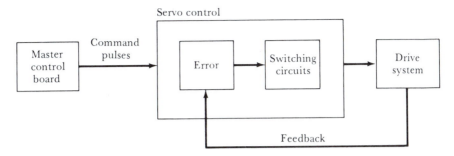

Figure 3–17 Block Diagram of the Servo Control Board

axis so that the repeatability and the accuracy of the robot are maintained.

The servo control board receives command pulses from the master control board. These pulses are then processed on the servo control board and transmitted to the axis motor. The axis motor has a feedback device that converts the rotational movement of the motor into electrical feedback pulses. These feedback pulses are transferred to the servo control board, where they are compared with the command pulses from the master control board. If there is a difference between the number of command pulses and the feedback pulses, an error signal is generated. This error signal is supplied to the axis motor, which establishes additional axis rotation.

Figure 3–17 is a block diagram of the components found on the servo control board. As the figure shows, the servo control board contains the servo amplifier circuitry that is responsible for the amplification of the *error signal*. Also found on the servo control board is a transistor *switching,* or silicon-controlled rectifier (SCR), *circuit* that is used to control the motor rotation. Chapter 8 describes in detail the operation of the servo system used in robotics.

INTERFACING BOARD

Interfacing is a very important task of medium- and high-technology controllers. The **interfacing board** is used for communication to peripheral devices, end effectors, and the controller's major components.

The interfacing board is the communication link established between the robot controller and peripheral devices. For example, an output signal from the master control board may be sent to the interfacing board to command the positioning table (the peripheral device) in an arc-welding system to rotate. The interfacing board may also be set up to receive input signals from peripheral devices and to send those signals to the master control board for action by the MPU.

Another interfacing process is the interfacing between the master control board and the end effector on the manipulator. For instance, a command signal generated on the master control board is routed through the interfacing board to tell the end effector to close its gripper. An input signal from the gripper is processed through the interfacing board to tell the master control board that the gripper is closed.

Many of the internal and external components use the interfacing board to communicate with the master control board. For example, data from the teach pendant is processed through the interfacing board to the master control board. Likewise, output signals from the master control board are processed through the interfacing board to the teach pendant.

The interfacing board is basically an input/output communication board. These input/output command signals processed by the interfacing board allow the controller to have total control over the entire work cell. Interfacing is discussed again in detail in Chapter 10.

SIGNAL PATH IN THE CONTROLLER

The **signal path** in the controller is very important to the operation of the robotic system. The sequence of events in the signal path is as follows: Signals generated from a source are processed by the controller, and controller commands then cause the axes of the robot to move. This path is developed in most high-technology controllers as described in the following paragraphs.

Figure 3–18A illustrates a block diagram of the signal path between the controller and the manipulator for a rotational movement. As the figure indicates, the signal path goes from the teach pendant, through the MPU and other devices, to the drive motor of the base. Notice that the controller contains most of the components discussed earlier in the chapter. The teach pendant is the device that provides the manual jogging of each of the various axes during the teaching process. The teach pendant is also used for programming the robot and for other operating characteristics of the system.

The controller is turned on by the operator after he or she checks the work cell of the robot to ensure that the job is ready to

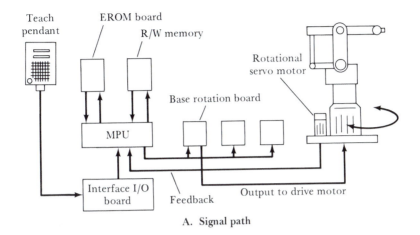

A. Signal path

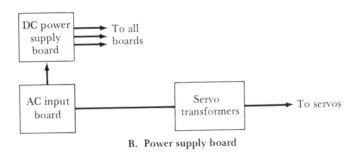

B. Power supply board

Figure 3–18 Signal Path for Driving a Rotational Axis

run and that no one is inside the work envelope pressing the on button. The 480 volts AC power is supplied to the controller's power supply. The AC input voltage is developed at the power input transformer and distributed to the servo transformers and the DC power supply board (see Figure 3–18B). Once the controller has made an electronic check of the internal circuits to ensure that there is no problem, the controller applies power to the servo motors. Now, the servo motors are ready to be commanded to move.

When the power is first applied to the controller, the controller does not know the manipulator's location in space. To develop this starting relationship, the operator must perform a zero-return operation. The **zero-return operation** calibrates the manipulator and the controller so that the controller has a reference point in space to operate from for positional data. This zero-return function is part of the control program that is stored in the executive memory section of the controller. Each time the zero-return function is executed, the manipulator will return to a certain position.

Figure 3–19 illustrates a zero-return point for the manipulator. In this example, the zero-return point is located −150° from the center of the manipulator's work envelope when an observer is looking down from the top of the work envelope. Once the zero-return operation has been completed, the manipulator is ready for program information.

The operator can now move the manipulator's axes into any position within the work envelope. The command signal from the teach pendant is sent to the controller through the teach pendant cable. The signal from the teach pendant is sent in a serial format, as shown in Figure 3–20. Notice that the first bit in the serial format is a *start bit*. The start bit lets the controller know that information is about to be sent over the cables. The 8 bits following the start bit are the *data command bits* of information. These command bits are sent to the MPU, asking the MPU to call up, from memory, the routine required for axis movement. The second-to-last bit is a *parity bit*, which ensures that all the correct data have been transmitted over the cable and that a false reading will not cause the wrong axis to be jogged. The final bit in the pattern is

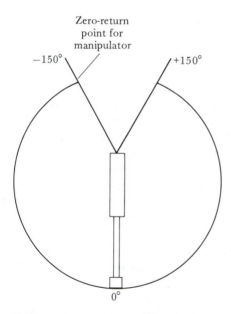

Figure 3–19 Zero-Return Position

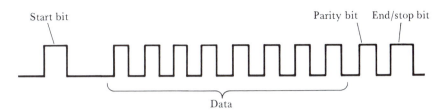

Figure 3–20   Serial Format for Data Transfer

the *stop bit,* which tells the controller that all the information has been sent.

The information is transferred on the input/output board. This board is used only for transferring data between the peripheral devices and the master board. Data for axis movement is then transferred to the master control board.

The positional data is now analyzed by the MPU. After analysis, the data is sent on a data bus to the output board and then to the servo drive system. The servo drive system sends command pulses to the axis drive motor, thus causing motion of the axis. As long as the operator is depressing the jog key for that axis, the axis will continue to move. Once the axis is in position, the operator releases the key, and the axis stops moving.

The operator of the robot can now program the positional information into the controller by depressing the programming key of the teach pendant. When the operator depresses this key, the current positional information is stored in the controller's memory. Thus, the operator can jog the manipulator to many different positions with the teach pendant. Once the manipulator is in position, the operator can program that point.

In many high-technology controllers, the operator can program up to 1200 different axis points. These points of positional information will remain in memory until the operator programs new data into the controller. Once this programming process has been completed, the operator can put the robot in a repeat mode, and the program will be played back. This program can be replayed as many times as desired.

≡ SUMMARY

The controller is the heart of the robotic system. It is used to store robot programs, control axis motion, and communicate with pe-

ripheral devices within the work cell of the robot. For these purposes, it contains several printed circuit boards that carry out these
various functions. General features of controllers include the microprocessing unit and memory boards, a teach pendant, a cathode
ray tube for display, feedback devices, and cooling devices.

The controllers used in the industry today are classified into
three areas: low-technology controllers, medium-technology controllers, and high-technology controllers. The low-technology controller is generally used in situations where limited axis motion is
required and limited control is needed over the entire work cell of
the robot. The low-technology controller is not easily reprogrammed. The medium-technology controller is used in areas where
reprogramming of the robot is important for its operation. The medium-technology controller can control from two to four axes of
movement with limited memory capacity. The high-technology controller has an expanded memory capacity and also is easily reprogrammed for different tasks. The high-technology controller provides
communication ports for external components within the robotic
work cell.

The controllers in the medium- and high-technology robots
employ the latest developments in electronic printed circuit boards.
The input power supply board is used to connect the plant's 480-
volt AC power supply to the operating system of the controller. The
input power supply board also contains the DC power supply board,
which converts the AC input voltage into DC operating voltage for
the controller.

The master control board houses the MPU and the memory
circuits. The control signals for communication with peripheral devices as well as command signals for axis motion are all generated
on the master control board.

The heart of the controller and of the master control board is
the microprocessor. In many medium-technology controllers, low-
powered, 8-bit processors are used. In high-technology controllers,
16- or 32-bit microprocessors are employed. This additional processing power allows the high-technology controller to develop more
computing power for overall control of the work cell. In both types
of controllers, the MPU has several components, including executive memory circuits and the central processing unit.

Memory capacity is very important in medium- and high-technology controllers. A memory board located within the controller
provides for program storage and for operation of the robot. The
memory circuits in the controllers are a combination of random-
access and read-only memory units.

The axis motion of the robot is very important when high levels
of repeatability and accuracy are needed. For the control of axis

motion, the controller has a servo control board. This board receives command signals from the master control board and transmits them to the axis motor.

The interfacing board is used to communicate with external components of the robot as well as components within the robot's work cell. The interfacing board is basically an input/output communication board.

Signal paths are identified in the controller for every operation. These paths are routed through the various printed circuit boards to establish communication with peripheral devices or to develop axis motion.

KEY TERMS

address bus
address latch
arithmetic logic unit (ALU)
bit
bubble memory
buses
byte
CAD/CAM
cathode ray tube (CRT)
central processing unit (CPU)
control program
data highway
digital codes
executive memory board
high-technology controllers
input/output (I/O) circuits
input power supply board
interfacing board
internal register (IR)
interrupt signal
I/O data latch

latches
low-technology controllers
master control board
medium-technology
  controllers
memory boards
memory map
microprocessing unit (MPU)
multi-tasking
nonvolatile memory
off-line programming
operational registers
positional data
program counter (PC)
read-write (R/W) memory
registers
servo-control boards
signal path
teach pendant
zero-return operation

QUESTIONS

1. How are low-technology controllers generally programmed?

2. What size of microprocessor is found in high-technology controllers?

3. Identify the five major sections of a high-technology con-troller. *Power, option, interface, servo-control, mp*

4. What is the range of memory size for many high-technology controllers? *32K — 1M*

5. Identify the type of link that is placed between the CAD system and the robot controller. *Outer Highway*

6. What circuit is responsible for converting the AC input voltage to pulsating DC voltage? *rectifiers*

7. Describe what the protection circuits protect the controller against. *high volt surges*

8. How is voltage from the DC power supply kept at a constant level? *regulators*

9. What is the main component on the master control board? *MPU*

10. What part of the MPU keeps track of the step of the program? *PC reg*

11. About how many bytes of memory are reserved for the MPU for storing data? *1 • 4K*

12. What section of memory is reserved for the parameters of the controller? *option register 4K*

13. What printed circuit board is located between the master control board and the axis motor? *axis drive servo control*

14. Which transformer is used to distribute AC voltage to the DC power supply board and the servo motors?

15. In what section of memory is the zero-return function stored? *exec. mem.*

16. Before one programs the robot, what is the first step to be performed? *Zero return function.*

17. Which bit is used to ensure that all the bits have been transmitted? *stop bit*

18. How is information transferred from the teach pendant to the controller? *pcc interface board*

19. Which board within the controller will not process axis motion commands? *interfacing*

20. What component allows the operator of the robotic system to move the axis to various positions within the work envelope? *teach Pendant*

4

Basic Robotic Programming

OBJECTIVES

Upon completing this chapter, you should be familiar with:

— Features of the user's program,
— Developing the program,
— Flowcharting the program,
— Machine coding the program.

INTRODUCTION

The robotic controller has two basic programs: the control, or executive, program and the user's program. The **control program** controls the basic operation of the robotic system. It is developed by the manufacturer, and it gives the controller the necessary routines for the robot's general operation. This program cannot be altered by the user without damage to the robotic operating system. The **user's program** is the one that the user writes to meet his or her applications. This chapter discusses the process for developing the user's program for medium- and high-technology controllers.

The development of the program for the robotic system involves four steps. The first step is defining the type of robot used. The second step is defining the task of the robot. The third step is identifying the sequence of events by which the program must be structured. The fourth step is identifying the conditions of the program.

The user's program is constructed in much the same way that a computer program is constructed. The program is developed through the four steps just outlined. Then, a flowchart is written to correspond to the program and to identify the programming process. Next, the flowchart is converted into machine codes, and, finally, the program is entered into the controller through a device for teaching the information to the robot. After the program has been entered, the program can be recalled from memory for replay of the stored steps.

This chapter describes most of the steps involved in constructing the user's program. The one step that is not described is the step of entering the program; this topic is discussed in Chapter 5. We begin our discussion here by looking at some features of the user's program.

FEATURES OF THE USER'S PROGRAM

The user's program is much like an ordinary computer program. A computer program develops a sequence of logical steps that will perform some type of operation. This operation might be adding two numbers or finding the correct mixture of gas and air in the carburetor of an automobile. The user's program has a similar process. The robot must be programmed to complete a task. The task could be welding a car's side body or palletizing boxes on an assembly line. Each of these applications requires a logical sequence of events.

In this section, the features of the basic components of the user's program are described. Also discussed are the features of programming languages and off-line programming.

Components of the Program

The user's program is basically like programs that are used with personal computers. That is, the program must have an identification number. The **program identification** provides convenient storage of and access to the program. This program identification could be a set of numbers, such as 1234, or might be the name of the program, such as "weld" or "palletize". When the user wants to recall the program, he or she simply gives the program number to the controller. Then, the program is loaded into the system for processing by the controller. Figure 4–1 illustrates some program numbers labeled by the letter R and followed by four numbers. The programs corresponding to these numbers are stored in the controller's memory.

The user's program must also contain positional data. *Positional data* describes the axis movement of the robot, and it can be given in one of two units of measure in the program. Axes that move in circular motion require units of degrees. Axes that travel in a linear pattern require units of inches or millimeters. Positional data must also be given for each axis of the robot. For example, for a five-axis articulate manipulator, each of the five axes requires positional data.

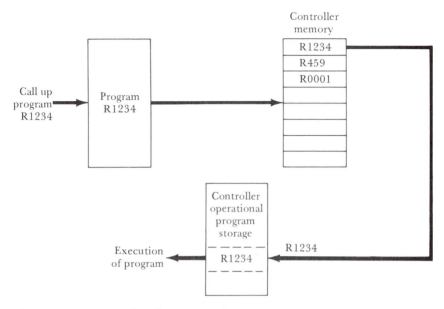

Figure 4–1 Program Identification Number

Figure 4–2A illustrates a typical block in a program that might be used for robot command data. The program block consists of the program number, that is, the letter R followed by four numbers. Also found in the block is the point data that identifies the positional data for each of the robot's axes. The positional data block has an address, labeled N000, which denotes the first set of positional data for the program. The address of the block is followed by positional data for each axis. The positional data is given for the R axis, the Z axis, the theta (T) axis, the alpha (A) axis, and the beta (B) axis. The R and Z axes are linear axes; therefore, the positional data is given in millimeters (mm in the figure) of travel. The theta, alpha, and beta axes are rotational axes, and their position is identified in angular position, or degrees of rotation. The second address block, N001, gives a similar set of positional data for a different manipulator position.

Figure 4–2B illustrates a typical five-axes articulate robot program. Again, the program number is the letter R followed by four numbers. The addresses of the various positional data blocks are N000, N001, and so on. The robot's positional data is given in angular degrees of operation for the five axes.

Another part of the user's program defines the **geometric moves**, that is, the movement that the robot's axes will make when it is moving to the desired position. The axes can be commanded to move

```
                                    R1234

    N000

          R 111.00 mm    Z 150.35 mm   T  147.00°
          A 101.00°       B  57.25°

    N001

          R 150.00 mm    Z 110.75 mm   T  90.00°
          A 101.00°       B  57.25°
```

A. Linear axes and angular rotation

```
                                    R1234

    N000

          T 110.00°   W  56.00°   U 00.00°
          A  90.00°   B 120.00°

    N001

          T  90.00°   W  75.3°    U 00.00°
          A  87.16°   B 120.00°
```

B. Axes for an articulate robot

Figure 4–2   Positional Data for a Robot's Axes

in a linear fashion or in a circular fashion. Each geometric move must be part of the program. In many applications, the special geometric moves of the axes are stored in the executive program of the robot. The user then only needs to place a special geometric code in the program to start the geometric moves.

Figure 4–3 illustrates the process of placing a linear geometric move in address N000. This address is labeled as P_1 in the figure. As the robot moves to the coordinate position labeled as P_2, the manipulator is commanded to move all of its axes in a linear path. The linear moves performed by the manipulator through such a program are very important in arc-welding, spot-welding, and painting applications.

The user's program must also define any special action that must be performed once the manipulator has reached the desired position. This action is called a **service request**. For example, a service request might tell the controller to start a welding process or to turn on a peripheral device.

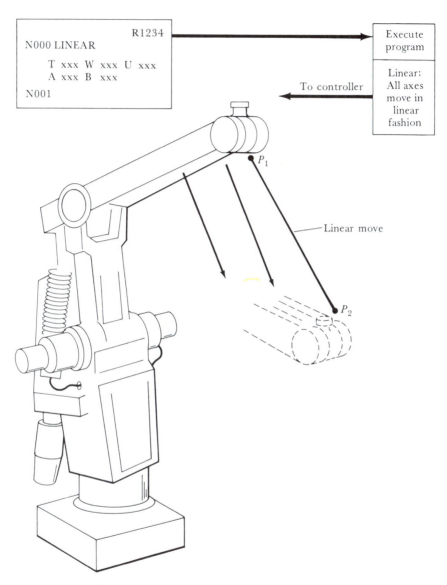

Figure 4–3 Robot Arm Making a Straight Linear Move

Service requests are part of the controller's executive program, but the programmer must use a special service code to call up the service requests to execute the different operations. For example, the command code CG1 might be used to tell the controller to close gripper 1 on the end effector of the robot. This command is stored in the executive memory of the controller. When this command is read in the program, the MPU calls up this information from memory so that it can execute the close gripper command.

The final component of the user's program is the **feed rate**, which is defined as the axis speed for reaching the desired position. The feed rate of an axis will vary according to the manufacturer's preprogrammed values. Also, the speed at which an axis moves can be either the composite speed or the single-axis speed. The **composite speed** is the speed of all of the axes moving together to reach a point. The problem here is that all of the axes can only move as fast as the slowest axis. In **single-axis speed**, the speed of that particular axis is the only limit.

The program containing the components just described must now be placed in a logical order by the programmer. The program is made up of many different addresses. But each of the addresses contains the same type of information, namely, a geometric code for axis movement, a feed rate defining the speed of the axis to a commanded position, the positional data, and a service request for the operation to be performed when the manipulator arrives at the desired location.

The number of programs that a controller can store depends on the size of user memory in the controller. In many high-technology controllers, user memory can range from 100 points to over 6000 points, where a **point of program information** is defined as an address of the program. Thus, for a 6000-point memory, only one program can be stored in memory if the program contains 6000 addresses of information. But three 2000-point programs can be stored in this memory.

Special Programming Languages

Many robotic manufacturers have developed their own programming languages that are to be used with their controllers. These programming languages are similar to the PASCAL computer language. For instance, the VAL language was introduced by Unimation; the AL language was introduced by Stanford University; the AML language was introduced by IBM; the HELP language was developed by GE; the MCL language was developed by McDonnell Douglas; the RAIL language was developed by Automatix; and the KAREL language was developed by GMF Robotics.

With these special programming languages, the robot does not require a special block within the program for developing axis movement. Instead, this movement can be programmed through a simple one-step command such as GOTO or MOVE. These commands allow the MPU to calculate the **approach vector**, that is, the path that the axis will travel to reach the point. The approach vector then directs the axis to move to the desired location.

Figure 4–4 illustrates a material-handling operation and the BASIC programming language. In Figure 4–4A, the various positional moves are outlined by the T values. Figure 4–4B shows the BASIC program; it gives the programmed points, the commands to the gripper, and the speed for the axes.

Figure 4–4A shows that the robot starts from point T5, which is called the start position. This position is given in the program of

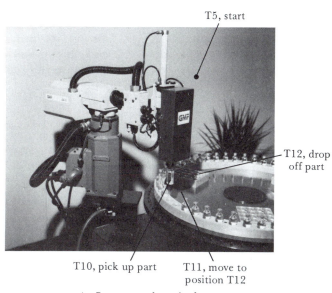

A. Programmed manipulator

```
 10   T5 = 0. 0. 15.0.
 20   MOVE T11
 30   WAIT + IE3:  SPEED 200
 40   OUTPUT +OGO 200
 50   MOVE T10 + T5:
 60   SPEED 10:  MOVE T10
 70   OUTPUT +OGI 200
 80   SPEED 150:  MOVE T10 + T5
 90   MOVE T11 T12
100   OUTPUT +OGO 200
110   GOTU 10
120   END
```

B. Program

Figure 4–4 Moving a Manipulator by Using BASIC Program Statements (Photo courtesy of GMFanuc Inc.)

Figure 4–4B in step 10. In step 10 of the program, notice that the positional data is given in Cartesian coordinates. In step 20 of the program, the manipulator is commanded to move to position T11. Step 30 commands the manipulator to wait and **index** the table— that is, to advance the table to the next position. The indexing command is given in step 30 by the + IE3 command. This step also contains the speed at which the robot should approach point T11.

Step 40 of the program commands the manipulator to open the gripper; the command for opening the gripper is the + OGO 200 command. Steps 50 and 60 command the manipulator to move to position T10 and pick up the part. Step 70 commands the manipulator to close the gripper; it uses the + OGI 200 statement. Step 80 is another movement command, and the manipulator is repositioned at point T11. Steps 90 and 100 position the manipulator over the drop-off pallet, and the part is placed onto the pallet with this routine. Step 100 again uses the open gripper command, + OGO 200. The program will recycle when statement 110 is read, and it will continue in this loop (steps 10–110).

Each time this cycle is completed, a register keeps track of the number of operations. This register-counting routine is a built-in program within the controller, and the count number can be inputted by the customer. In the example of Figure 4–4, the cycle will end after 100 bottles have been placed onto the pallet. At this time, the program will end, using step 120.

Off-Line Programming

With the use of the special programming languages, off-line programming can be integrated into the robot's operation. Off-line programming, as described here, is programming written at a computer terminal. Since all the moves of the robot can be programmed through special language commands, the programmer can sit at a computer terminal and program the manipulator's moves. The program can then be transferred to the controller's memory through a data bus and executed. In this situation, the controller memory can be reduced in size, because the main computer memory holds the program.

Off-line programming also uses the CAD/CAM system. The movements developed by the manipulator can be drawn on the CRT of the terminal, and the programmer can then regulate the manipulator's moves through the CRT. Again, the information can be downloaded to the controller for execution by the manipulator. Generally, when the program has been downloaded, the operator of the robotic work cell must go to the shop floor and touch up the positional data of the program. Thus, the operator of the robot is a very important part in the operation of the system.

Figure 4–5 illustrates a typical connection between the off-line programming terminal and the robotic work cell. Notice that several robot controllers can be connected to the system through the data bus. Hence, one programmer can develop programs for an entire assembly line.

DEVELOPING THE PROGRAM

The robotic program must contain the information necessary for the total operation of the work area. Therefore, the programmer of the robot must take several steps into consideration when developing the robotic program. As mentioned in the Introduction, these four steps are defining the type of robot, defining the task of the robot, identifying the sequence of events in the program, and identifying the conditions of the program. Each step of development is discussed in the following subsections.

Defining the Type of Robot

In many cases, the robot will already have been selected. So, the programmer's task now is to develop a program that will execute the defined task of the robot. To do so, the programmer must have knowledge of the type of robot, whether the robot is, say, a cylindrical coordinate robot or an articulate robot. The programmer must also know the number of axes and the amount of reach on those

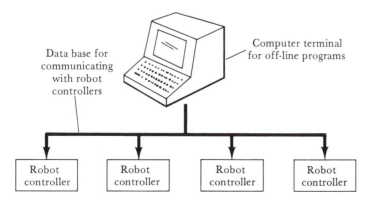

Figure 4–5 Downloading from a Computer Terminal to the Robot Controllers

axes. In addition, the programmer must have information about the work cell of the robot.

The controller and its amount of user's memory space are two further important concerns. The number of points that can be put in memory is limited by the available memory space. The programmer must also understand the language used by the controller and must be able to write commands for axis movement, service requests, and geometric moves.

Defining the Robot's Task

The second step in developing a successful program is defining the task the robot manipulator will perform. Will the robot have to lift parts? Will the robot spot-weld? Will the robot be inspecting various parts? All the tasks the robot is to perform should be specified clearly before the programmer proceeds to the third step.

Identifying the Sequence of Events

The third step in developing the program is identifying and listing the sequence of events that must occur if the robot is to perform its task. For example, the task might require the robot to start from a *perch position,* that is, from a resting point of the manipulator, as shown in Figure 4–6. All the manipulator's moves—and, hence, the program steps—will be executed from this point.

In listing the sequence of events, the programmer must know the controller's responsibilities, such as whether it must control various peripheral devices. For instance, the controller might send an output signal to a conveyor line to move a part in front of the manipulator, as shown in Figure 4–7. In this situation, the controller must also receive an input feedback signal from the conveyor line that the part is in position for the manipulator to pick it up. The controller is expecting these signals because of programming statements. Without special commands in the program, the controller will ignore any input signals, which we might think of as service requests. In the programming task, the programmer must also be aware that it takes time for the requests that come from peripheral devices to be received by the controller.

Another important concern of the programmer is the cycle time of the program. Many times, production schedules are determined by the company's industrial engineers. They have calculated the amount of time needed for the conveyor to move into location, for the robot to grip the part, and for the part to be moved by the

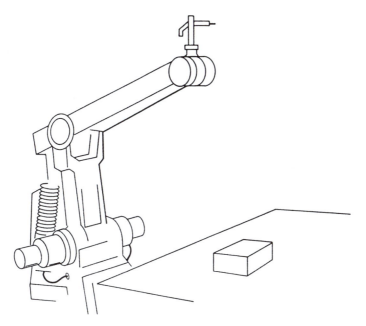

Figure 4–6   Manipulator in a Perch Position

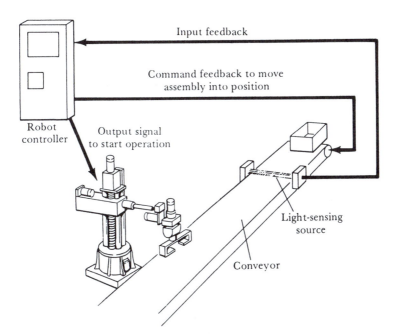

Figure 4–7   Input/Output Operation

manipulator. The programmer must allow for these times in the program so that the robot can meet the cycle time demands.

In addition to considering cycle time, the programmer must take into account the time required to open and close the manipulator's hand grippers. The program should contain delay times that will allow the robot to grasp the part. If these delays are not programmed, the controller will give the manipulator only enough time to sense the presence of the part.

Identifying the Conditions of the Program

The fourth step in developing a program is to identify the conditions, or limits, that are placed in the program by the executive program of the controller. These limits include the speed at which an axis will accelerate and decelerate and the maximum stroke length of each of the axes.

The conditions also identify the options that are available to the programmer. For example, if the robot's task is to stack parts on a pallet—a process called *palletizing*—the programmer must program each point of the task. Thus, if there were 100 parts to be palletized, then the programmer would have to program all 100 points. But in many high-technology controllers, palletizing programs are already written into the executive memory. Thus, in this case, the programmer need only call out the special palletizing routine, enter the number of parts to be palletized, and enter the number of stacks and rows of parts. The controller's memory will then compute the rest of the information and palletize the parts.

Figure 4–8 illustrates a manipulator palletizing finished parts onto an indexing table. The manipulator stacks the parts two high. Once the parts are stacked, the manipulator's controller commands the indexing table to move to the next open station. The program routine in the controller commands the manipulator to stack all the parts on the bottom row first and then to stack the parts on the top row.

FLOWCHARTING THE PROGRAM

To develop a smoothly operating program, the programmer should develop a **flowchart**, or a step-by-step diagram, of the program before the program is written and entered into the controller. In this section, we will examine the terms, symbols, functions, and operations involved in flowcharting.

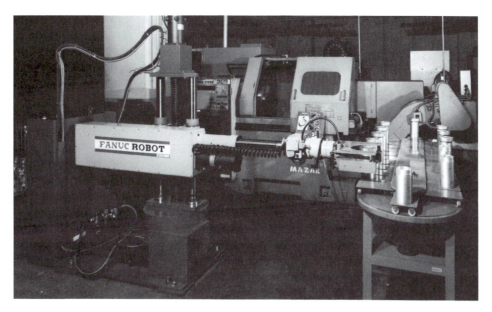

Figure 4–8   Palletizing Parts on an Indexing Table (Courtesy of GMFanuc Inc.)

Flowchart Symbols

The standard flowchart symbols are illustrated in Figure 4–9. These basic flowchart symbols can be found in any introductory programming book.

Figure 4–9A illustrates the oval shape, which is used as a **terminal symbol**. This symbol identifies a stop, a start, or a point in the program where there is an interruption. Figure 4–9B illustrates the **input/output symbol**. This parallelogram is used to indicate input or output data or a control signal.

Figure 4–9C shows the **decision block**. This diamond shape indicates the location in the program where a decision will be made on the basis of input or output data from a peripheral device or from the contents of the program. Figure 4–9D shows the **process block**. Generally, this rectangle indicates that some process is taking place in the controller. For example, a process block is used to indicate that the axes are moving to a point location.

Figure 4–9E illustrates the **node symbol**. The node point in the program is the point where signals are coming in from different locations for processing.

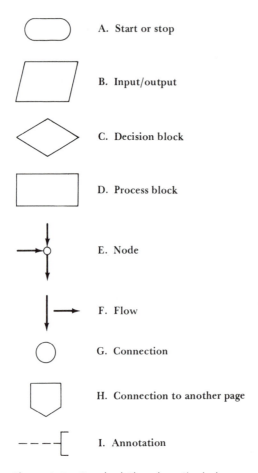

A. Start or stop

B. Input/output

C. Decision block

D. Process block

E. Node

F. Flow

G. Connection

H. Connection to another page

I. Annotation

Figure 4–9 Standard Flowchart Symbols

Figure 4–9F shows the **flow symbols**. Flow symbols, or arrows, are used to connect blocks in the program's flowchart. These symbols also show the general direction of program development.

Figure 4–9G illustrates the **connection symbol** used in long flowcharts. For many programs, the flowchart might have to be tied together at several points. The connector symbol identifies these connection points. Figure 4–9H illustrates the **off-page connection symbol**, which is used when the flowchart goes to another page.

The **annotation symbol** is shown in Figure 4–9I. This symbol is used, for example, when the programmer assigns machine codes to the flowchart. For instance, suppose that, in the process block, the programmer indicates that the axes will move quickly toward the programmed location. The annotation symbol can then be used to show that the speed is a linear rate of, say, 1000 millimeters per second.

The flowchart is used primarily to organize the programmer's thoughts about various movements and events in the program. After the programmer has developed the flowchart, the operation of the program can be related to a smooth cycling time. Several flowcharts are discussed in the following sections. These flowcharts show how the flowchart symbols might be used in some typical examples of robot operation.

Sample Flowchart

Figure 4–10 illustrates a sample work cell and robotic tasks for which a programmer might develop a flowchart and a program. The moves that must be programmed can first be developed off-line and then placed into the controller's memory for operation.

As indicated in the figure, the manipulator moves from the home position to position 1 in a linear path. From position 1, the manipulator moves all of its axes in a simultaneous fashion to position 2, right above the part; this move is accomplished with a fast axis speed. This fast motion will decrease the cycle time of the robotic operation. From position 2, the manipulator then moves

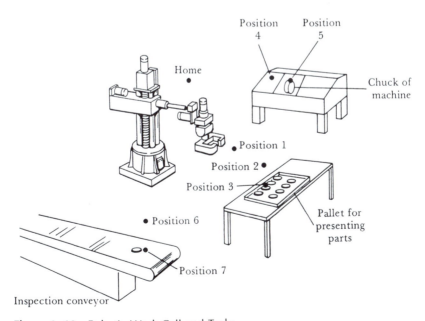

Figure 4–10 Robotic Work Cell and Tasks

slowly in a linear fashion down onto the part. This slow move allows the sensor in the gripper to sense whether a part is present; it also keeps the manipulator from crashing into the part.

The part is picked up at position 3, and the manipulator returns to position 2. From here, the manipulator is commanded to move at a high rate of speed to position 4, in front of the machine chuck. The robot manipulator then moves slowly into the chuck of the machine, to position 5. The manipulator moves back to position 4 outside the machine to wait for the machine operation to be completed. At the completion of the machine operation, the manipulator moves into the machine to pick up the finished part. Next, the manipulator moves back from the machine to position 4 and moves to position 6 above the conveyor. Now, the manipulator places the finished part on the conveyor (position 7) and returns to position 6. From this location, the manipulator moves to position 1 to start the cycle over again.

The movements illustrated in Figure 4–10 are simple moves for the robot and are not difficult to program. The flowchart explaining the robot's movements for this example is shown in Figure 4–11. Notice that the flowchart begins with the terminal symbol. The next blocks in the flowchart are process blocks, where the programmer indicates four items of information. These four lines are the geometry required by the axes so that the robot moves to the programmed position; the velocity of the axes (remember that velocity identifies how fast the axes will move); the positional data for the program; and the service request.

Notice that each programmed move has a separate process block (and address number) that includes the four items of information. Thus, each block describes the various tasks that must be performed by the robot. That is, a geometric command must be executed by the controller. This command describes how the robot's axes will reach the programmed point. The process block also identifies the speed at which the axes will travel in order to reach that point. For a linear axis, this speed will always be given in millimeters per second; for a rotational axis, the speed will be given in degrees per second.

The position that all of the axes must reach is the next information given in the block. This positional information gives the coordinates of the axis for the point in space. The final command in the process block is the service request. The service request describes the opening or closing of the gripper, the welding being performed, or any special operation performed by the end effector. These components of the flowchart and the program will aid the programmer later when the machine code must be matched with the process blocks.

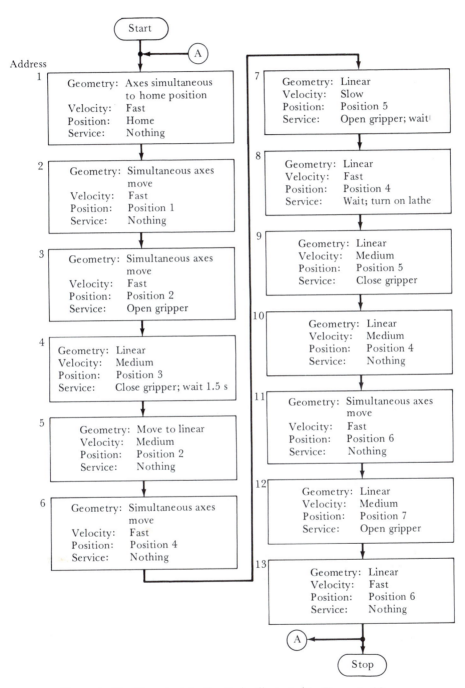

Figure 4–11 Flowchart for the Tasks Illustrated in Figure 4–10

The flowchart continues for several blocks, and then the connector symbol is used. This symbol indicates that the program continues at another location. A good habit in flowcharting is to place an identification number or letter in the symbol, which allows another user to find the other end of the flowchart. The flowchart continues for several additional steps, and then it is terminated.

The flowchart illustrated in Figure 4–11 is for a single-block type of program. That is, after the program has reached the terminal ending block, the program is over. The operator of the work cell has to recycle the operation if the robot is to repeat the task. This manual recycling defeats the purpose of the robot—to work without human supervision. In the next section, we discuss a method—namely, branching the program—that avoids the problem of manual recycling.

Branching the Program

Figure 4–12 illustrates the **branching** programming process; branching allows the program to be self-operating. The branch in a flowchart indicates that when the program reaches a certain point, the program will automatically go (branch) to another specified location in the program. The branching operation works well for the example of Figure 4–11 because it is only a one-cycle operation.

In Figure 4–12, notice the addition of a decision block in the flowchart. The decision block (Is pallet empty?) allows the program to branch to another part of the program. For the example of Figure 4–10, this branching operation allows the robot to return to the pallet, position 1, and pick up another part at the end of the cycle. That is, the robot will return to position 1 and repeat the entire operation. The branching continues until the program is stopped by the operator.

In flowcharting and programming, there are two types of branching: the conditional branch and the unconditional branch. In the **conditional branch**, a condition for branching is set by a peripheral device that sends a signal back to the controller or by a comparison made in the controller's memory. For example, suppose that when the pallet is empty (Figure 4–10), the programmer wants the robot to automatically stop. In this case, the programmer must write a conditional branch that involves the empty pallet. Figure 4–13 illustrates this conditional branch. Notice that the branch has returned the program—and thus the robot's manipulator—to the home position. In the flowchart of Figure 4–12, the process that sends the program back to the starting point is an **unconditional branch**.

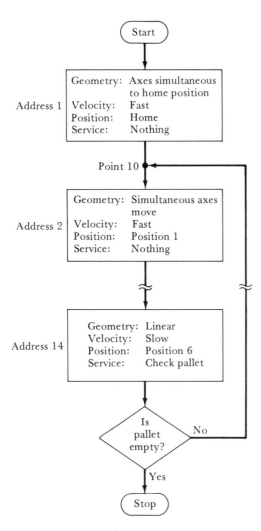

Figure 4–12 Branching Process

Input and Output Signals

An important part of the total operation of the robot is the control exercised by the controller. This control can be developed from the input/output signals. Thus, during certain operations in the program, the programmer might use input or output signals from the peripheral equipment as controlling devices. From information received from the peripheral devices, the program will branch to other locations.

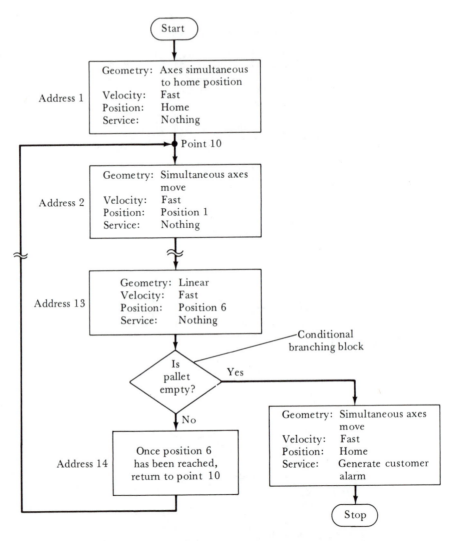

Figure 4–13 Conditional Branching

In the original example of Figure 4–10, the robot loads the finished machined parts on an inspection conveyor. At this point in the program, the programmer might want the table to advance automatically to the next open location—that is, the programmer might want to have indexing of the table. Indexing the table allows the robot to place the part onto an open pallet.

Figure 4–14 illustrates a numeric keypad assembly operation with a vision system. In this assembly operation, four keys from a feeder tray are placed on a keypad located on the conveyor line in

Figure 4–14   Keypad Assembly System (Courtesy of Automatix Inc.)

front of the manipulator. A camera is mounted on the manipulator for the purpose of determining the orientation of the keypad on the conveyor line. Input data received from the camera causes the axes of the manipulator to be offset to compensate for the position of the keypad. So that all of the various moves are made for a correct assembly operation, the programmer should write a flowchart using input and output signals to ensure that the program controls the entire assembly operation.

Figure 4–15 illustrates a flowchart for this assembly operation for one row of keys. In this flowchart, the program starts by moving the manipulator to position 1, which is the position over the feeder tray. At this time, the vision system identifies whether parts are present in the feeder tray. If no part is present, the robot waits at position 1 and turns on an alarm to alert the operator that no part is present. If parts are detected by the vision system, the manipulator is commanded to move slowly into the feeder tray and pick up four of the keys with its gripper.

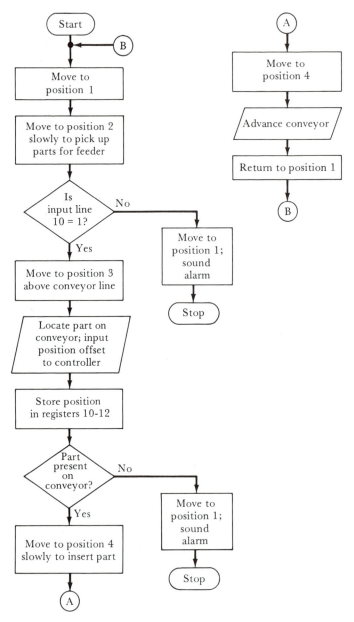

Figure 4–15 Flowchart for the Keypad Assembly System

Next, the robot leaves position 2 and, at a medium speed, approaches position 3, which is the position over the conveyor line. Again, the vision system is used to detect whether a keypad is on the conveyor line. If a keypad is on the line, the vision system senses the orientation of the part. If the part is offset, then the offset data is stored in registers 10–12. This offset data is read by the controller. Then, the controller adjusts the position of the manipulator as it moves to position 4. Here, the robot inserts the keys into the keypad. At this time in the program, the vision system is also looking for the presence of parts on the conveyor line. If a part is not on the line, the robot moves to position 1 and generates an alarm. The alarm must be reset by the operator.

After the insertion of the keys, the manipulator is commanded to move to position 1. At that time, the conveyor drive system receives a signal from the controller to advance to the next position. This process continues over and over again during the assembly operation.

The keypad assembly operation requires that five rows of keys be assembled in the same fashion. The flowchart in Figure 4–15 illustrates the assembly operation for only the last row.

Register Operation

As stated in Chapter 3, the controller's MPU contains certain registers. These registers are electronic storage areas. In a program, the registers can be used to store the total number of cycles completed in an hour by the robotic cell, for example, or the register can be used to store the total number of parts on a pallet.

The programmer can use the MPU register conditions as branching conditions. Figure 4–16 illustrates a flowchart for a register-controlled program. The program for the flowchart executes ten times and then returns to position 1 and waits.

The first step of the flowchart is the start block. Once the program starts, the robot moves to position 1, and the program automatically clears register 1, which is the register that will be used in this operation. In the clearing operation, the number placed in register 1 is zero. Now, the robot moves to position 2 and then to position 3. In this block of the flowchart, notice that the programmer has indicated that register 1 is used to keep track of cycles of operation. At position 3, the program will increment register 1; that is, a 1 will be added to the contents of the register. Since the register was cleared to zero, the new number in register 1 at this time is 1.

The robot next moves in a linear fashion to position 4 and checks to see whether input signal 5 has turned on. A decision must be made at this point. If input signal 5 is on, then the work cell has an empty pallet. In this case, the robot moves to position 5 and the program stops. If input signal 5 is not on, the program continues and checks register 1 to see whether it is equal to 10. If register 1

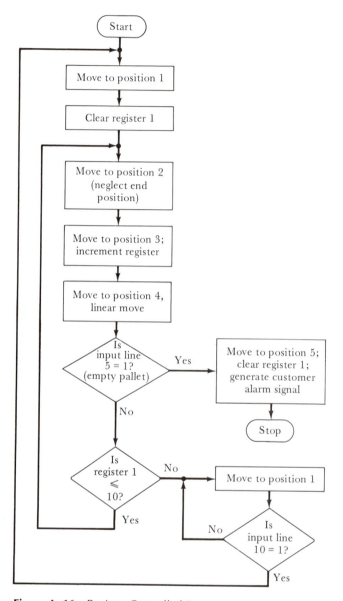

Figure 4–16 Register-Controlled Program

is not equal to 10—that is, is less than 10—the program branches back to the start and runs again.

The program continues in this fashion until register 1 is equal to 10. When register 1 is equal to 10, the manipulator moves to position 1. Another decision block is placed after this move in the program. Input line 10 is checked to see whether it is high (that is, is equal to 1), meaning that an empty pallet has been replaced with a full pallet. If a full pallet is not in place, the manipulator waits at position 1 until the full-pallet condition is met.

In this example, the register is incremented every time the cycle is complete. However, register operation can also be decremented. That is, the programmer can load a 10 into register 1 and, each time a cycle is complete, have a 1 *subtracted* from the register.

Also notice that two branches are used in the flowchart of Figure 4–16. The flowchart has a main branch that returns the program to position 1. Each time the program returns to position 1, though, register 1 is cleared, and the register will never be allowed to count up to 10. The minor branch in the program allows the operation to branch to position 2, avoiding the clearing operation.

Timer Operations

Many times, the programmer must provide delay times in the program to allow all operations to be completed. This timing condition in the program is called the **null condition**. In a flowchart, the null condition is represented by a decision block, as shown in Figure 4–17. The null condition means that the robot will perform no operation while the program is waiting. The program waits for an input signal to turn on before advancing to the next program step.

Skip Function

Another common programming operation is the skip function. The **skip function** allows the programmer to skip an operation and go to the next operation in the program. Figure 4–18 illustrates the flowchart symbols for a skip function. The figure shows two pro-

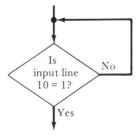

Figure 4–17   Null Condition

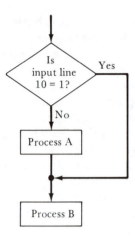

Figure 4–18   Skip Function

cesses, process A and process B. The skip function allows the pro-grammer to branch around process A if an input signal turns on. If the input signal does not turn on, the program will execute process A and then process B.

IF–THEN–ELSE Statement

The **IF–THEN–ELSE statement** allows the robot (and the program) to perform process A until a given input is developed by a peripheral device and sent to the controller. Figure 4–19 shows the flowchart blocks for this statement. Once the input signal is read by the

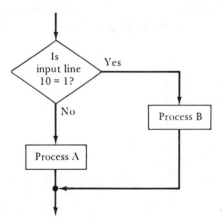

Figure 4–19   IF–THEN–ELSE

controller, the program jumps around process A, performs process B, and completes the task developed by the program. Once process A has been completed, the program loops back to the start and performs the operation over again.

DO WHILE Statement

The **DO WHILE statement** is a very powerful command in programming. This command allows the program to execute a process or another program while the program is waiting for an input signal from a peripheral component. Figure 4–20 illustrates the flowchart symbols. Notice that the program will execute process A until the input signal is turned on. Once the input signal is turned on, the process branches to another location.

Main Programs and Subprograms

Many applications require the programmer to develop several small programs in a single program. Such a program might be needed, for example, to perform spot welding on a car moving down an assembly line. The main program allows the robot to read the body style code of the car. The main-line program then branches to another program; the program it branches to depends on the contents of the body style code.

The additional programs within the main program are called **subprograms**. In the example of a welding operation just discussed, the subprogram will contain the spot-welding information, such as the path the robot must follow. Once the spot-welding program has

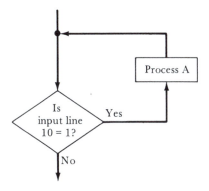

Figure 4–20   DO WHILE

been completed, the subprogram returns to the main-line program until another body style code is read.

Figure 4–21 illustrates a flowchart for a main-line program and several subprograms. Notice that, at the end of the main-line program, it will branch to one of several different subprograms. The GOTO SUB PROG command calls up the subprogram for the body style that is present. This body style code must be read several steps ahead in order for the program to call up the subprogram. Also notice that the main-line program has a fail-safe operation.

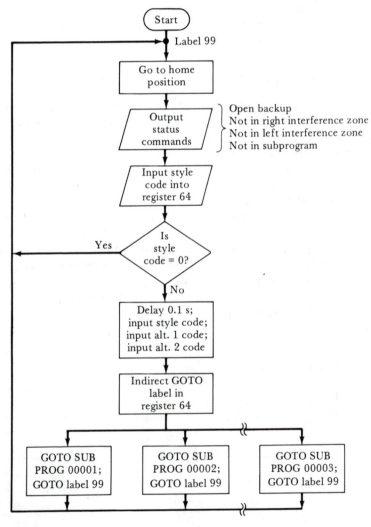

Figure 4–21 Flowchart for a Main Program and Subprograms

That is, if the body style code is not read, then the program branches back to the beginning of the main-line program, thus causing the manipulator to return to the home position.

MACHINE CODING THE PROGRAM

Machine coding is the process of converting the flowchart and program concepts into machine codes. The machine codes describe to the controller the various movements, the velocity, and the programmed position of the manipulator. All of these commands are placed in a special machine code that is different for each manufacturer's controller. Since the coding for each controller is different, this text will not discuss machine codes.

SUMMARY

Robotic programs fall into two categories: the control program and the user's program. The control program is the executive program developed by the manufacturer for the general operation of the robot. The user's program is the one that the user writes, and it contains positional data, axis velocity data, geometric axis moves, and the service requests once the manipulator is at the programmed location.

Many manufacturers have developed their own programming language. Each of these high-level languages allows axis movement within the limits of the program and the limits of the controller. Some of these high-level languages are similar to PASCAL.

The program for the robot can be written off-line, that is, at a computer terminal away from the shop floor. The program can then be downloaded to the controller through a data bus.

Developing a robotics user's program consists of four steps: defining the type of robot, defining the robot's task, identifying the sequence of events, and identifying the conditions of the program. By considering these four steps, the programmer ensures that the program contains the information necessary for the total robotic operation.

A flowchart helps the programmer develop the program. The flowchart lists the various steps that take place in the program. The steps of the flowchart describe all the various moves that the robot must make in order to complete the task. These steps also show the assignments of all the input and output commands the controller needs in order to manage the work cell.

In the construction of a flowchart, several basic flowchart symbols are used. For example, the terminal symbol is an oval that identifies the starting point or the stopping point of the program. The parallelogram is the symbol used to identify input or output signals from the program. These input or output signals, for example, might be used to move a conveyor line.

The decision block is a diamond-shaped symbol. It is used in the flowchart to identify whether a condition has been met in the program. Lines with arrowheads are flow symbols, and they show the direction of the flow of the program. Arrows that point into or out of a circle show node connection points. These arrows are used when branching statements are employed in the program and the program must return to a specific location.

Many times, one part of a flowchart must be connected to another part on a different page. This connection is shown by a circle with a letter or number placed inside the circle. The symbol that is shaped like a home plate in baseball is called the off-page connection. If the flowchart is lengthy and requires several pages, the off-page connection symbol is employed. The annotation symbol is used to remind the programmer what is meant by the shorthand that might be placed in the process blocks or the decision blocks.

Part of the controller's memory contains registers. The programmer can use the registers to store data during program operation. The registers can be used to keep a parts count during an operation of the manipulator. The register is controlled through the program.

Many different functions can be employed in the flowchart operation. Some of these functions are the skip function, the timer function, the DO WHILE function, and the IF–THEN–ELSE function. Each of these functions help the programmer to develop routines that allow the robot to perform the many simple and complex tasks required by the job.

In many large programs, the programmer may need to develop subprograms that operate from a main program. For example, the main program may develop the basic motions that move the manipulator into a perch position. Then, on the basis of an input signal from an external component, the program branches to a subprogram. The subprogram generally contains all the moves that the robot must make in order to perform its task.

KEY TERMS

annotation symbol   branching
approach vector   composite speed

conditional branch
connection symbol
control program
decision block
DO WHILE statement
feed rate
flowchart
flow symbols
geometric moves
identification number
IF–THEN–ELSE statement
indexing
input/output symbol

machine coding
node symbol
null condition
off-page connection symbol
point of program information
process block
service request
single-axis speed
skip function
subprogram
terminal symbol
unconditional branch
user's program

QUESTIONS

1. How are different robotic programs separated in the controller? *By add a uses prog. bt by name · prog ID*

2. Positional data for a linear axis is generally given in what units of measurement? *mm/inches*

3. Positional data for an angular axis is generally given in what units of measurement? *degrees*

4. What part of the program's address identifies the motion path that the axis or axes will travel? *-*

5. What is the speed of all the axes moving together called? *Compact*

6. Speeds for linear axes are generally given in what units of measurement? *cm/sec*

download from — central compule 7. How can the controller's physical memory size be reduced?

8. See Figure 4–5. At what speed will the axes approach point T10? *speed 10*

9. In the special programming languages, what one-step command develops axis movement? *GoTo n move*

10. What is the final step that is used when one is programming on a CAD terminal?

11. What is the first step in developing a user's program?

12. Refer to Figure 4–7. How does the robot know that the part is in position? *sensing device*

13. What information must be available to implement a palletizing routine? *# of Parts; # of Stacks & Rows.*

14. Draw the flowchart symbol used for input/output signals.

15. Draw the flowchart symbol used for branch statements.

16. What type of branching is used for conditions set by peripheral devices? *Conditional Branch*

17. Give an example of the use of input or output signals in a program. *input that parts are present output to sound alarm*

18. Name one typical use for registers in a program. *To keep count*

19. Name three functions that can be done to manipulate registers in program operation. *load add subtract*

20. What condition allows the manipulator to wait at a certain location? – *Waiting for a part on a conveyer belt.*

21. Draw a sample flowchart for the skip function, and explain the operation.

22. When is the IF–THEN–ELSE statement used in programming applications?

23. What is the name given to the process of converting the flowchart into robot language commands? – *Machine Code*

24. Write a flowchart for a manipulator that moves to five positions; it picks up a part at position 2 and deposits the part at position 4. This process is to be done ten times and then stopped, at which time the manipulator will move to position 1 and a customer alarm will be activated.

25. What command should always be placed in the first step of the program when the program uses registers? *Clear the reg.*

(22) *when a condition exist (input high) a process will be done – when the input is not true then another process it done.*

5

Operational Aids

OBJECTIVES

Upon completing this chapter, you should be familiar with:

— Teach pendant,
— Operator's panel,
— Manual data input panel,
— Computer control.

INTRODUCTION

The robot controller has several external components that are necessary for the operation of the system during programming and for the daily operation of the robotic cell. These components are the teach pendant, the operator's panel, the manual data input panel, and the computer control. Each component is described in this chapter.

The teach pendant allows the programmer to enter positional data and to jog the axes to programmed positions. The manual data input panel aids the programmer in inputting data and in editing the programmed information. The operator's panel allows the operator to control the entire system. Computer control allows the programmer to program the manipulator by using a CAD system or an engineering workstation. In addition, through computer control, the programmer can obtain information about the daily production of a robot or whether the robot has broken down and is in need of repair.

TEACH PENDANT

Figure 5–1A illustrates the major components of the teach pendant. The teach pendant is generally employed with medium- and high-technology controllers. The **teach pendant** allows the programmer to enter positional information into the user's memory of the controller. It also is used to jog the robot's axes into position.

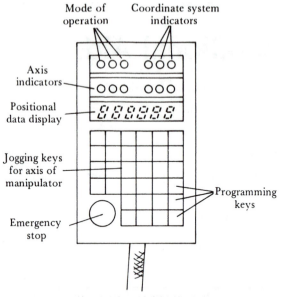

Mode of operation

Coordinate system indicators

Axis indicators

Positional data display

Jogging keys for axis of manipulator

Emergency stop

Programming keys

A. Components of the teach pendant

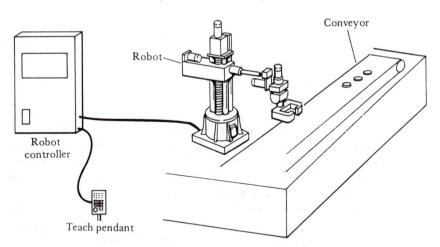

Conveyor

Robot

Robot controller

Teach pendant

B. Teach pendant brought into the work envelope

Figure 5–1 Teach Pendant

In Figure 5–1A, notice that the teach pendant has several indicators. These indicators display the mode of operation of the controller, the axes coordination, and the positional data. The teach pendant also has axis jogging keys, programming buttons, a dead-man switch, and an emergency stop button. Each of these components of the teach pendant is described in the following subsections.

The teach pendant is used to teach the controller the various positional points of data. It is a portable component that can be carried into the work envelope by the robot's programmer, as indicated in Figure 5–1B. All of the information required by the programmer can be viewed with the teach pendant.

Modes of Operation

Generally, the robot's controller has three basic **modes of operation**: the teach mode, the test mode, and the repeat mode.

During the teaching of the program and the inputting of positional data, the robot is placed in the **teach mode**. This mode allows the operator of the system to jog the axes of the manipulator to the various positions as detailed on the programming sheet. The programmer has full control over each axis movement from the teach pendant. That is, the programmer can jog an axis up or down or rotate an axis a full 300°. The programmer can also control the speed of axis movement. Generally, when programming the robot, the programmer will set the speed of the axis to the lowest possible range, for safety of the robot.

In the **test mode** of operation, the programmer can repeat the taught paths of the robot. The test mode allows the programmer to replay these positional points without turning on any of the output signals. During the test mode, the operator should be clear of the manipulator's work envelope. This will ensure safe operation of the manipulator. In the test mode, the programmer can generally go through the program in a single-step process, allowing the programmer to see the exact points that were programmed.

An example of the test mode of operation is illustrated in Figure 5–2. For this figure, the programmer has just completed the teaching process for an arc-welding operation. All of the feed rates, the geometric moves, the positional data, and the service codes have been programmed. So, before the programmer starts actual production, he or she will play back the program in the test mode. All of the axis movements are completed during this mode, and the programmer can determine whether the taught paths will give the production arc weld the best cycle time and the best weld. Also, if the programmer feels that certain points in the program need to be adjusted, the program can be edited through the use of the teach pendant. Finally, during this testing, the test mode lamp is illuminated while a test weld is being made to show that the operation is being performed. No welding arc is generated during the test mode, however.

Once the programmer is satisfied with the taught paths from the test mode, the robot can be placed into the repeat mode. The

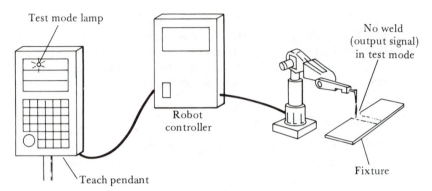

Figure 5–2   Test Mode with a Teach Pendant

repeat mode of operation puts the robot into production runs. Positional path information that has been programmed can now be placed in the repeat mode. These paths will continue to run until the program encounters a stop signal or an alarm condition.

Coordinate System and Axis Indicators

Figure 5–1A shows the location of the coordinate system indicator lamps. The lamp for the **coordinate system indicator** turns on when the programmer selects a system in which the robot manipulator is to be jogged. For example, if the robot is to be jogged in articulate coordinates, the programmer selects this mode by pressing the articulate coordinate selection key on the teach pendant. Placing the robot in the articulate coordinate mode causes all of the axes of the manipulator to move to the desired location.

The coordinate system indicator lamps stay illuminated as long as the programmer remains in that coordinate system. Changing to another coordinate system changes the indicator display.

The lamp for the **axis indicator** works in exactly the same way as the lamp for the coordinate system indicator. The lamp turns on when the programmer selects the axis to be jogged. The lamp remains on as long as the programmer is jogging the selected axes.

Jogging the Axes

The teach pendant has keys that allow the programmer to specify movements of the axes. These keys are called the **jogging**, or positional, **keys**. The programmer may also position the robot's axes through the use of joysticks or the lead-through-teaching method. Each of these three methods is described next.

Jogging Keys. Figure 5–3 illustrates a typical teach pendant used for jogging the axis of the robot into position. Notice that on this particular teach pendant there are ten jogging keys used for directionalizing axis motion. When the programmer uses these keys, the robot can be moved from the minimum to the maximum of the axis stroke limits.

The top jogging keys are the X axis keys. The X axis key on the left is used for an X axis movement to the left. The X axis key on the right is used for an X axis movement to the right. The Y axis keys are located right below the X axis keys. These keys, when depressed, cause axis motion in a diagonal, up or down direction. The vertical upward and downward directions of the axes are controlled by the Z axis, up and down keys.

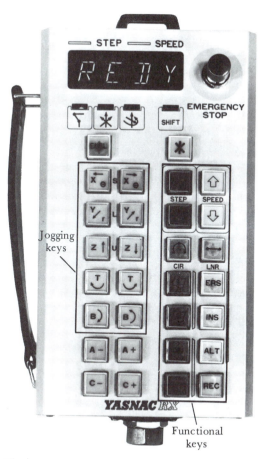

Figure 5–3 Jogging Keys on a Teach Pendant (Courtesy of Hobart Brothers Company)

Rotation around the base is controlled by the next set of keys, the T (theta) keys. The T axis keys allow the manipulator to rotate around its base. The base rotation is controlled in a clockwise or counterclockwise direction. The final set of keys is the set of B (beta) keys. These keys are used for the direction of rotation of the wrist. The wrist rotation can be upward or downward.

As an example, suppose that the programmer is working with a robot in the Cartesian coordinate system. On the teach pendant, the programmer uses the X, Y, and Z axes keys for jogging the robot's axes. These keys are arranged so that movement of the axes can be in a positive or a negative direction in the Cartesian coordinate system (as indicated by the arrows on the keys in Figure 5–3).

The Cartesian mode of operation allows the programmer to automatically jog the robot along the Cartesian coordinates of the workpiece. The steps that the programmer must take are as follows: First, the Cartesian coordinates of the robot are aligned with the Cartesian coordinates of the workpiece. Next, these coordinate values are placed in the program register for later recall for reference positioning. Programmed points are then entered as the programmer jogs the robot along these coordinate values.

Once the program has been written and entered, the programmer can replay the programmed movements. In the replay mode, the robot will move according to the programmed points.

At some time during a robot's operation, the axis limits might be overextended. In this case, the robot manipulator is in the condition called overtravel. The **overtravel condition** results when the programmer jogs the axis of the robot beyond the limits of the stroke. For detection of overtravel, the robot has overtravel limit switches mounted on the mechanical unit, as shown in Figure 5–4. A cam is generally used to detect contact between the axis and the overtravel switch. When contact is made, the robot automatically goes into an alarm condition and ceases movement of the axis. The programmer will then have to reset the alarm of the overtravel switch, jog the axis of the robot off the overtravel condition, and continue with the program. Resetting the alarm condition can be done with the teach pendant.

Joystick. Figure 5–5 illustrates a teach pendant that includes a joystick. The joystick is located in the upper right-hand corner of the teach pendant. The **joystick** is used to position the various axes of the manipulator into position over the part. The joystick's positioning corresponds directly to the axes' positioning.

Notice also in Figure 5–5 that this teach pendant has various *numeric keys;* these keys are used during programming to place

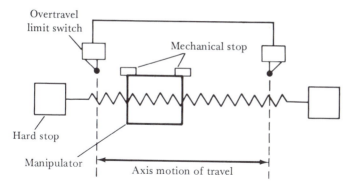

Figure 5–4 Overtravel Switches

numeric values into the program. To the left of the numeric keys, in the two bottom rows, are special command or **function keys** that open or close the gripper of the manipulator. Above the four gripper keys are the *coordinate jogging keys*. These keys allow the programmer to select the coordinate system for jogging the manipulator.

This teach pendant also has a seven-segment display that shows the various steps of the program. These steps are shown in the top row of the display. The bottom row of the display illustrates the

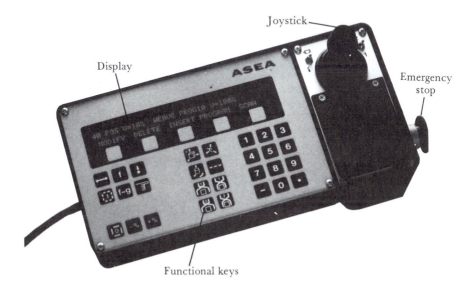

Figure 5–5 Joystick Mounted on a Teach Pendant (Courtesy of ASEA Robotics Inc.)

functions of the soft keys. The term *soft key* means that these no-
tations can be changed. The **soft keys** put the controller into a selected mode of operation. For example, if the programmer wants to
select a certain displayed position as the programmed position, then
the programmer presses the PROGRAM soft key, and the positional
information is programmed.

The button on the far right of this teach pendant is the emergency stop button. This button causes the manipulator's axes to
come to a fast stop.

Lead-through-Teaching Method. Movement of the axes may also
be accomplished by the **lead-through-teaching process**. No jogging
keys are used with this method. Instead, the programmer leads the
tip of the manipulator or the tool to the various desired positions.
Once the manipulator is in position, the data is recorded as positional data.

In Figure 5–6, for example, the programmer has the manipulator stationed at point 1. This position is recorded in the program.
The programmer then leads the tip of the tool to point 2 and records
that position. The path that the manipulator takes from point 1 to
point 2 is recorded in the controller's memory. The lead-through-
teaching process is used with many medium-technology controllers.

Programming

As stated earlier, the teach pendant is a portable device that the
programmer can carry into the work envelope of the robot. With
the teach pendant, the programmer can guide the axes into the
positions required to complete the task. The programmer can then

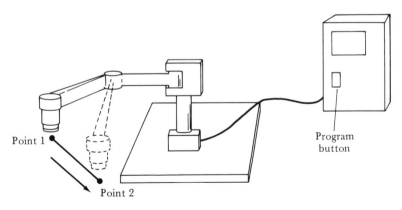

Figure 5–6 Lead-through-Teaching Method

record the necessary positional information with the teach pendant. Medium- and high-technology controllers allow the programmer to program positional information, feed rates, geometric moves, and service requests with the teach pendant.

Figure 5–7 illustrates another type of teach pendant that might be found in an industrial robotic system. Three important **programming keys** are located in the second row: the delete, input, and program keys. The **delete key** allows the programmer to delete (remove) any address of information from the program. For example, suppose that, during the programming of the manipulator's position, an error in positional data was inputted. By using the delete key, the programmer can remove this data. The key directly to the left of the delete key is the **input key**. The input key is used to insert program information into the program. The key to the left of the input key is the **program key**. This key is used for programming

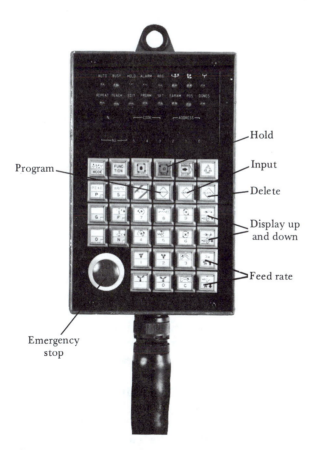

Figure 5–7   Programming Keys (Courtesy of GMFanuc Inc.)

commands into the robot's address block. For example, suppose that the robot is at a desired location, and a geometric code needs to be programmed. By using the programming key, the programmer can place this geometric code into the program.

The keys located in the fifth and sixth rows at the far right are the **feed rate keys**. These keys are used to input the rate at which an axis will travel to a desired position. The feed rate keys can be used, for instance, during the operation of a robot in order to increase or decrease the speed at which the manipulator travels to its position.

The two keys above the feed rate keys are the **display-up** and **display-down keys**. The programmer uses these keys to call up different addresses in the program. For example, if the programmer wants to view the positional data at address 1, and the program is at address 3, then the programmer merely depresses the display-down key to reach address 1. These keys allow the programmer to command all the robot's work directly from the teach pendant without returning to the controller.

Safety Buttons

So that the programmer has complete control over axis movement when he or she is in the work envelope of the robot, three safety buttons are included on the teach pendant. These buttons are the hold key, the deadman switch, and the emergency stop button.

The **hold key** is used to stop all axis motion of the manipulator. In Figure 5–7, the hold key is located in the top row, the third key from the right. When this key is depressed, it remains detented. To start manipulator motion, the programmer must release the hold key and depress the start button to begin the program.

The hold key on the teach pendant allows a programmer who is in the work envelope of the manipulator to halt all axis motion. But, generally, the hold key only holds the axes in position. Power from the main input line keeps the axes alive. At any time, the axes can start moving if the robot is ordered to move by someone at a remote location. If the programmer is in the work envelope at the time, then the programmer may be injured.

The **deadman switch** is used to ensure that the robot will stop if the teach pendant is dropped by the operator. The deadman switch is usually located on the teach pendant handle so that the operator must depress the deadman switch to hold the teach pendant.

The **emergency stop button** completely stops the robot's movement. It automatically applies brakes to the axes, causing the axes to stop and hold their positions. The emergency stop button also causes the power to the axis motor to be removed by breaking the input power line from the AC power of the plant. Figure 5–7 illustrates the location of the emergency stop button on the teach pendant. The button is always within reach of a programmer who is in the work envelope of the robot.

OPERATOR'S PANEL

Medium- and high-technology controllers always have some type of control panel that allows the operator of the robotic work cell to maintain control of the robot. The panel is called the **operator's panel**.

The operator's panel is often located on the robot's controller and is used by the daily operator of the system to keep control of the manipulator and controller. It allows the operator to stop the manipulator's axes in an emergency or to put the axes in the hold mode. It also allows the operator to turn control of the robot's controller over to a programmable controller or to protect the memory of the controller.

Figure 5–8A illustrates a typical operator's panel found on a controller. The operator's panel contains the different controls needed for operation of the manipulator.

Figure 5–8B illustrates the details of the operator's panel. The **start button** is used to start the daily program once it has been called from memory. Next to the start button is the hold button, which allows the operator to stop axis motion at various locations. Once the hold button has been depressed, the operator can restart the operation of the program by depressing the start button. Both the start button and the hold button have indicator lamps. The indicator lamp turns on when the robot is in the start (or the hold) mode. This lamp will illuminate when the button is depressed from the teach pendant or from the operator's panel.

The **manual/automatic switch** is used after the program has been called up and the robot has been returned to its zero position. Once the start button has been depressed, the robot will begin program operation. The operator can now put the robot in automatic operation by using the automatic switch. The programmer can also put the robot in remote operation by using the **remote switch**. In remote operation, the robot can be controlled from a control panel elsewhere in the plant or from a programmable controller.

The **memory protect switch** on the operator's panel keeps the programs in the controller from being overwritten or erased. The emergency stop button serves the same function as the emergency stop button on the teach pendant. It causes all axes to stop moving, and, at the same time, it removes the power from the motors that

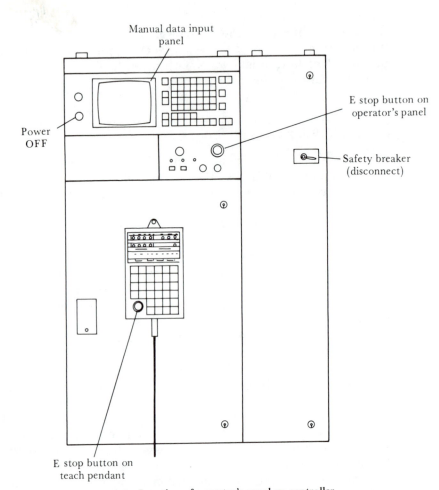

Manual data input
panel

Power
OFF

E stop button on
operator's panel

Safety breaker
(disconnect)

E stop button on
teach pendant

A. Location of operator's panel on controller

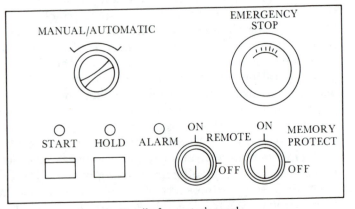

MANUAL/AUTOMATIC

EMERGENCY
STOP

START   HOLD   ALARM   ON   REMOTE   ON   MEMORY
                                  OFF   PROTECT   OFF

B. Detail of operator's panel

Figure 5–8   Operator's Panel

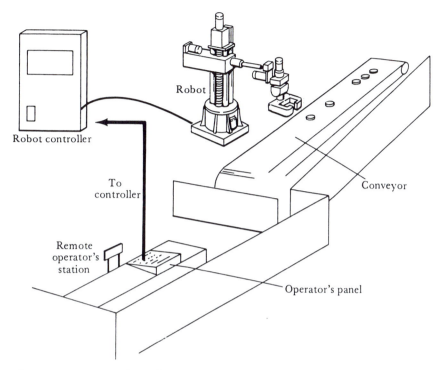

Figure 5–9 Remote Operation

drive the various axes. Before returning the robot to normal operation, the operator must adjust the work envelope to eliminate the condition that caused the operation to stop.

Figure 5–9 illustrates a typical remote operation for a robot. The operator's panel usually found on the controller is duplicated at a remote location. In this way, an operator can control several robots from one central plant location. All of the components on the remote operator's panel are the same as the components on the operator's panel of the controller.

Note that if the remote switch on the operator's panel of the controller is activated, then the remote operator's panel takes control of the system. Quite often, the remote switch on the controller is left on. Thus, when a programmer wants to change programs, he or she cannot do so. Control of the system must be returned from the remote location to the controller in this situation. In order to ensure the safety of the operator, the emergency stop button always has the highest control. This means that, if the control is at a remote location, the operator can still stop the robot at the emergency stop located on the teach pendant.

MANUAL DATA INPUT PANEL

The **manual data input (MDI) panel** is used to input data to the program, to input data to program control registers, or to edit program information. Figure 5–8A illustrates the location of the manual data input panel on a controller. The manual data input panel generally includes an alphanumeric keyboard, a cathode ray tube (CRT) for display of various program functions, and the power on/off buttons.

Figure 5–10 illustrates the details of a typical MDI panel and the keys found on the panel. The CRT is used mainly for display of program information. For example, by depressing the PARAM (parameter) button, the programmer or the operator of the system can call up, from memory locations, the various operating parameters of the controller and view them on the CRT. Then, the programmer, through the editing process, can change the parameters of the system to meet the changing needs of the program.

Programs that have been entered from the teach pendant can also be displayed on the CRT of the MDI panel. The entire program block can be displayed, thus allowing the programmer to take a look at the total program. Positional data, program addresses, geometric move codes, feed rate codes, and service request codes can all be viewed on the CRT. Any program information can be changed through the use of the visual display and the editing procedures.

The mode key on the MDI panel allows the programmer to change the modes of the controller from the MDI panel. With this key, the operator can change the controller from, say, the teach mode to the repeat mode of operation. The **cursor keys** allow the

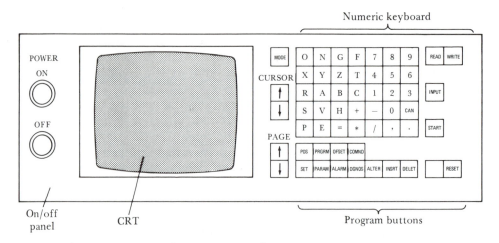

Figure 5–10 Manual Data Input Panel

cursor display on the CRT to be moved to a variety of locations on the screen. The **page-up** and **page-down keys** are used to display an entire new page of data. For instance, if the programmer wants to view the program at address 12, then by pressing the page-up or page-down key, the programmer can reach that location.

Figure 5–10 also illustrates the **alphanumeric keyboard** of the MDI panel. This keyboard is used to input data (numbers and letters) from a location other than the teach pendant. For example, if the programmer wishes to place the code G01 into the program, then this sequence is pressed on the alphanumeric keyboard. The keys on the MDI panel match the keys on the teach pendant. Thus, all the operations that can be performed from the MDI panel can also be accomplished from the teach pendant. The major difference between the MDI panel and the teach pendant is that the axes cannot be jogged from the MDI panel.

The **read/write keys** on the keyboard allow the programmer to read or write data into external memory devices. The input key operates in the same manner as the input key on the teach pendant. If data must be added to certain locations of the program, the input key allows the programmer to insert this data. For instance, once the G01 code has been inputted via the alphanumeric keyboard, then the programmer depresses the input key to insert this code into the program. The input key is also used to insert data into the different operational registers of the program. The start key is used to start the operation of the program.

The final block of keys on the MDI panel is the block of programming keys. These keys are used to call up different screens to the CRT. For instance, if the programmer wants to call up the program screen, then the PRGRM key is depressed. The programmer can then work with that screen for entering programs or editing programs. For example, during the programming of the manipulator, data might have been used in the program that will not perform the task. The program data shown on the CRT can be changed through the DELET, ALTER, or INSRT programming keys.

Like the operator's panel, the MDI panel may also be a remote operating device, allowing it to be used with several robot controllers. Thus, the programmer can carry the MDI panel to different robot locations. With remote operation, only one MDI panel will be required to program many controllers in a plant.

COMPUTER CONTROL

Many new industrial plants around the world are developing a totally automated process. In this process, the operation of the plant is controlled from one central computer. The computer is networked throughout the plant to the different machining operations, some of which may be robotic cells.

Computer control in robotics is off-line programming through a CAD system or engineering workstations. The programmer can control and program all of the robots from one central computer location. All of the control paths, all of the various registers, and all of the various codes for service, geometric moves, and feed rates can be taught, programmed, and changed from the main computer terminal.

The total operation involving computer control of an industrial plant is called a **flexible manufacturing system (FMS)**. Figure 5–11 illustrates a typical FMS work cell. The major control in this operation is the programmer who sits at the CAD terminal. This terminal for the main computer stores all of the control programs for the CAD system and the CAM operations. From this terminal, the programmer can cause the robot to move to the various positions on the conveyor line. Each of the different robots in the FMS will have a special task that it must perform. This task is programmed and downloaded to each robot controller. As the parts pass down the assembly line, each manipulator will perform some task on the assembly of the part.

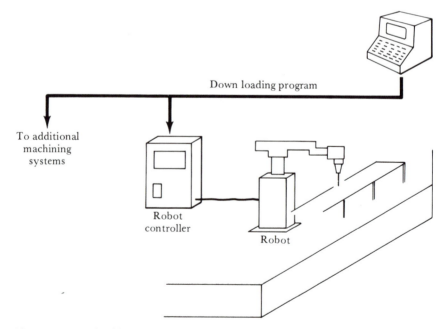

Figure 5–11   Flexible CAD/CAM Manufacturing System

In the FMS, the programmer of the work cell sits in the middle of the assembly operation to control all of the various operations of the manipulators. Feeder lines are also controlled through the CAD system to ensure that the correct number of parts is flowing onto the final assembly line.

The arrows in Figure 5–11 illustrate an important feature of the FMS: the communication link between the main computer and the work cells on the plant floor. The following subsections discuss this communication link and the advantages of the FMS.

Communicating with the Cells

One of the most difficult processes in the CAD/CAM operation is establishing the communication link between the main CAD computer terminal and the robots on the plant floor. For this operation to work, all of the various components must speak the same language. Thus, all of the work cells in Figure 5–11 must communicate with the main computer. So, all of the work cells must have a protocol language that will allow communications to take place with the host computer within the plant.

Figure 5–12 illustrates a typical hierarchy of communications within the plant for an FMS system. Notice that each of the components within the cell are connected to computer interface units (CIUs). These CIUs are then connected to a microcomputer cell

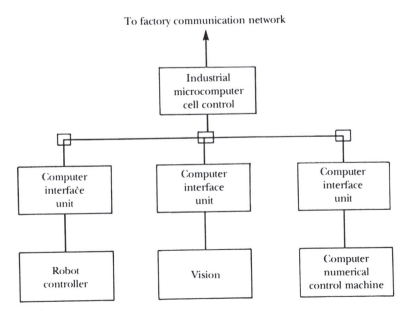

Figure 5–12　Basic Cell Controller Network

controller, and finally to a plant level computer. The plant level computer gives the instructions to the cell controller. The controller then commands the various components within the cell to operate. Data from the robot controller, vision, and the Computer Numerical Machine can communicate back and forth with the plant's host computer. Many cells, such as the one shown in Figure 5–12, can be networked together to receive and carry out instructions from the host computer.

Since each manufacturer has its own programming language, the communication process can be very difficult. But this communication problem can be solved by interfacing the work cells with **programmable controllers (PCs)**. The PC interfaces with the computer systems and establishes the necessary communication links. The communication link can be implemented by using the standard interfaces of RS–232, RS–422, half duplex, full duplex, and a standard baud rate.

Advantages of the Flexible Manufacturing System

Automated factories have certain advantages over traditional factories with stand-alone, single work cells. For example, the programmer can serve many different robotic cells at one time from the CAD terminal. Also, different programs from the operation can be stored in one central location rather than on separate floppy disks, where they might get damaged in the production environment.

Another advantage of the FMS is that it allows the programmer to upload programs from the robot to the main computer and to download programs from the computer to the robotic work cell. This system, though, has a serious security consideration. A password has to be incorporated in the program so that only authorized programmers of the system have access to the programs.

The FMS also provides a communication link to notify the operator if troubles develop in the program. Through the main computer, the operator can communicate that information to the programmer. The programmer can then, through the uploading process, call up the program to the CAD terminal and make the necessary modifications to the program.

With the FMS, the main computer keeps track of the different applications that the work cell has performed. Within an instant, the main computer can request the work cell to give a report on production for the past hour of operation. Through the main computer, the operator can find out the actual position of the servo motors of the manipulator, the current position in the program, or the positioning of the various axes.

The FMS, as you may have suspected, is an expensive operation for an industrial plant. But the communication link between the main CAD computer and the work cell allows production to keep up with demand. So, in the long run, FMS will save money for a plant.

SUMMARY

The robot controller has several aids that can be used during programming and operation. These aids are the teach pendant, the operator's panel, the manual data input panel, and computer control.

The teach pendant is used to program and teach positional information for the manipulator. The teach pendant is portable and can be carried into the work envelope. The programmer can then jog the axes of the robot into program positions. This positional data is recorded and stored for later recall.

The operator's panel is a built-in device of the controller as well as a separate device in a remote location. The operator's panel gives the operator full control over the robotic work cell. Various switches on the operator's panel provide control of the daily operations of the controller and the manipulator.

The manual data input panel is used to edit program positional locations or to edit registers that have direct control over the program. All of the operations that can be performed with the teach pendant can also be performed with the MDI panel, with the exception of jogging of the axes.

The flexible manufacturing system is used in many automated factories. The FMS allows control of the entire manufacturing operation to be handled through a main computer and a CAD terminal. From the main computer, the programmer can program the different robots in the FMS operation through one CAD terminal. The main computer communicates to the CAD terminal and the work cells through various communications networking systems. The major advantage of the FMS is that only a few individuals are needed to keep track of the entire plant's operation from one central location.

KEY TERMS

alphanumeric keyboard
axis indicator
coordinate system indicator

cursor keys
deadman switch
delete key

display-up and display-down keys
emergency stop button
feed rate keys
flexible manufacturing system (FMS)
function keys
hold key
input key
jogging keys
joystick
lead-through-teaching process
manual/automatic switch
manual data input (MDI) panel
memory protect switch

modes of operation
operator's panel
overtravel condition
page-up and page-down keys
program key
programmable controller (PC)
programming keys
read/write keys
remote switch
repeat mode
soft keys
start button
teach mode
teach pendant
test mode

QUESTIONS

1. Name three displays found on the teach pendant.

2. Which mode of operation allows the manipulator to be jogged into position and that position recorded?

3. Which mode of operation allows the manipulator to move to all the programmed positions at programmed speeds and to execute all service requests?

4. Which set of keys on the teach pendant allows the axes to be positioned?

5. Describe the condition of overtravel.

6. What function does the joystick serve on the teach pendant?

7. What two safety features are found on the teach pendant and the operator's panel?

8. Which safety feature removes power from the axis drive system?

9. Where is the operator's panel located?

10. What is the MDI panel used for during the programming mode?

11. What component of the MDI panel displays program information?

12. What system can be used for off-line programming applications?

13. Describe the key component of the flexible manufacturing system.

14. What component is used to interface the computer with the machining operation on the plant floor? *CNC comp interface unit*

15. Why are soft keys used on the teach pendant? *To put the controller in a selected modes*

16. Which keys on the teach pendant adjust the speed of axis travel? *feed rate*

17. Which teach pendant keys allow the programmer to view various addresses of the program? *keep / down*

18. Describe the lead-through-teaching method of programming the manipulator. *move the man. from pt - pt.*

19. Which component can be used to reset an overtravel alarm? *teach pendant*

20. Where is the on/off button located on the controller? *jogging*

21. Which key on the teach pendant causes the base of the manipulator to move in a clockwise or counterclockwise direction? *T Key*

22. After releasing the hold key, what should the programmer do next to start motion of the manipulator? *depress the start key.*

6

Hydraulic and Pneumatic Drive Systems

OBJECTIVES

Upon completing this chapter, you should be familiar with:

— Principles of hydraulics,
— Hydraulic system symbols,
— Hydraulic actuators,
— Directional controls,
— Hydraulic pumps,
— Hydraulic reservoirs,
— Pneumatic systems.

INTRODUCTION

Robotic manipulators may use one of three basic drive systems: electric, hydraulic, or pneumatic. As noted earlier, the weight to be picked up by the manipulator may be an indication of the type of drive system employed. In this chapter, the fundamentals of hydraulics and of pneumatics will be explored. The hydraulic system uses liquids under pressure, while the pneumatic system uses air under pressure. Each system has components that distribute this pressure to the various actuators that perform the work of the robotic system.

The major part of this chapter deals with the principles of hydraulics and the various components needed in hydraulic systems. The final section of the chapter discusses pneumatic systems, whose principles of operation are quite similar to the principles of hydraulic systems.

PRINCIPLES OF HYDRAULICS

The word *hydraulics* is derived from the Greek word for water. Therefore, the study of **hydraulics** can be considered to be the study

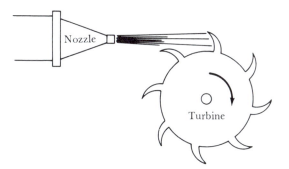

Figure 6–1 Water from a Nozzle Used to Turn a Turbine

of water flow. Figure 6–1 illustrates one way that water can be put to work. Water is directed from the nozzle to a turbine that has blades. As the water strikes the blades, it causes the turbine to turn. In this case, the energy from the flowing water is transferred to the rotary motion of the turning turbine.

However, the hydraulic systems in robotic manipulator drives do not use water; rather, they use oil. This oil is placed under pressure so that the energy from the oil is transferred to the movement of the manipulator. **Pressure**, or force, is the amount of push that is applied on a given area. In terms of measurement, pressure is the amount of push per unit of area of the surface acted upon, and it is usually measured in pounds per square inch (abbreviated lb/in², or psi). The pressure in a hydraulic system is developed from the push applied to a confined fluid.

The difference between the water that moves the turbine and the pressurized liquid is simply in terminology. The energy developed from water flowing from the end of a nozzle is called **hydrodynamic energy**. In contrast, the pressure of the fluid that causes the movement of the manipulator's arms is called **hydrostatic energy**.

In the following subsections, we will explore some basic principles of hydraulics: pressure, hydraulic flow, pressure drop, and Bernoulli's principle.

Pressure

Pressure is developed in a system when the fluid used encounters some type of opposition. The pressure can be developed in two ways: through the use of a pump and through the use of a weight placed on the fluid. An example of each method is shown in Figure 6–2.

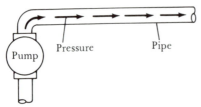

A. Use of pump

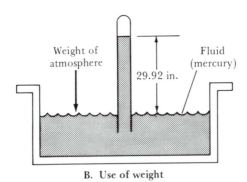

B. Use of weight

Figure 6–2   Developing a Pressure on Fluids

In Figure 6–2A, a pump is used to pass fluid through a pipe. Because the pipe is small, it creates pressure in the fluid line. In Figure 6–2B, a weight is placed on the fluid. This weight is our own atmosphere, which applies a pressure of 14.7 pounds per square inch to the fluid—in this case, the fluid is mercury. The atmospheric pressure forces the mercury up the tube to a height of 29.92 inches. This height results at sea level; other heights result when the location is above or below sea level.

Hydraulic Flow

The pressure developed in the robotic hydraulic system transforms the hydraulic energy into movement of the manipulator. This movement is caused by the flow of fluid through the various pipes in the hydraulic system. Thus, **hydraulic flow** in the system gives the actuator its motion. This flow is developed by a pump.

Flow in a hydraulic system is measured in two ways: as velocity of the fluid and as flow rate of the fluid. **Velocity of a fluid** is the average speed at which the fluid's particles pass a given point, or the average distance the particles travel per unit of time. The measurement of velocity is either in feet per minute (fpm), in feet per second (fps), or in inches per second (ips).

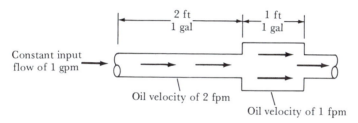

Figure 6–3 Oil Velocity as a Function of the Cross-Sectional Area of a Pipe

Flow rate of a fluid is a measure of the volume of a fluid passing a point in a given time. Large volumes are measured in gallons per minute (gpm). Small volumes are measured in cubic inches per minute (in³/min).

Figure 6–3 illustrates the difference between velocity and flow rate. As shown in the figure, 2 feet of the small pipe are needed to hold 1 gallon of oil, but only 1 foot of the large pipe is needed to hold 1 gallon of oil. With a constant flow rate of 1 gallon per minute, the oil travels 2 feet per minute in the small pipe. But when the oil reaches the large pipe, it must travel only 1 foot per minute. Thus, under the constant flow rate of 1 gallon per minute, the velocity of the oil either increases or decreases, depending on the cross-sectional area of the pipe.

Flow and Pressure Drop

Whenever a fluid is flowing, there must be a condition of unbalance to cause the motion. Therefore, when fluid is flowing through a pipe, the pressure is always greatest at the point closest to the input of the fluid. As the distance from the input increases, the pressure decreases.

Figure 6–4 illustrates this decreasing pressure. The pressure at the input, or the head point, is the greatest (maximum) pressure. As the fluid flows through the pipe, the friction in the pipe causes the pressure to decrease. These pressure decreases are shown by the different levels of fluid in the tubes along the pipe's path. The differences in pressure along the pipe are called **pressure drops**.

Bernoulli's Principle

Hydraulic fluid has two types of energy: kinetic energy and potential energy. **Kinetic energy** develops from the weight and the velocity of the fluid. **Potential energy** develops from the pressure of the fluid.

Daniel Bernoulli, a Swiss scientist, found that when a hydraulic system has a constant flow rate of fluid, energy is converted

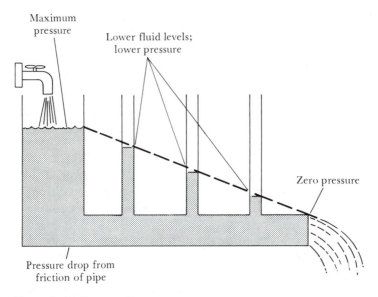

Figure 6–4   Pressure Drop

from one form to the other each time the cross-sectional area of the pipe is changed. **Bernoulli's principle** says that the sums of the potential (pressure) energy and the kinetic energy at various points in a system must be constant if the flow rate is constant. And as we saw earlier, when the pipe's diameter changes, the velocity changes. Thus, the kinetic energy, which depends on velocity, either increases or decreases.

Figure 6–5 illustrates Bernoulli's principle. In the small pipe at the left, pipe velocity—and hence kinetic energy—of the fluid is maximum. Therefore, since the kinetic energy is maximum, the

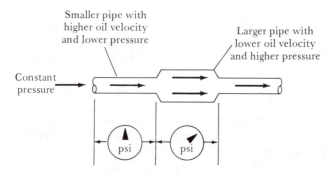

Figure 6–5   Bernoulli's Principle

pressure (potential energy) is low. The velocity and kinetic energy decrease as the fluid passes through the large pipe. The kinetic energy loss, according to Bernoulli's principle, is made up by an increase in pressure. As the fluid passes through the small pipe to the right, the pressure drops to the value it had in the small pipe at the left.

HYDRAULIC SYSTEM SYMBOLS

Hydraulic drive systems and circuits are of various types. So, symbols are used to identify the different devices. The hydraulic circuit is much like an electric circuit, which uses different symbols to represent the different components. A hydraulic circuit has two basic components: lines and rotational components. Each type is discussed in the following subsections.

Lines

Figure 6–6 shows a simple hydraulic circuit with the different components used in hydraulic operation. The hydraulic pipes and tubes for fluid passage—which are called **hydraulic lines**—are drawn as single lines in the schematic. There are three types of lines used in hydraulic systems: the pressure line, the pilot line, and the drain line.

The **pressure line** is shown as a solid line in Figure 6–6. This line carries the main stream or flow of fluid. The pressure line has three parts: the pump inlet line (or suction line to the pump), the main pressure line, and the return line to the tank.

The **pilot line** is shown as a long dashed line in Figure 6–6. This line controls operational valves or other components in the hydraulic system.

The **drain line** is illustrated by a short dashed line in Figure 6–6. The main function of this line is to carry leakage oil back to the storage area, which is called a *reservoir*.

The rectangular symbols at the ends of the lines in Figure 6–6 are the reservoir symbols. An open symbol, like the symbols in the figure, means that the reservoir is vented. If the symbol used is a closed box, it means that the reservoir is pressurized. When the lines going into the reservoir are drawn to the bottom of the symbol, the notation means that the input lines terminate below the fluid level.

The *relief valve* shown in Figure 6–6 is used as an outlet when the pressure in the system rises above the desired level. As long as the pressure remains below this level, then the relief valve is closed.

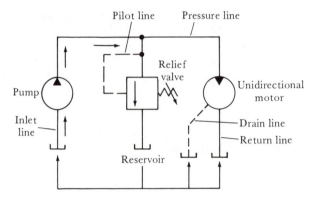

Figure 6–6   Schematic Diagram for Hydraulic Lines

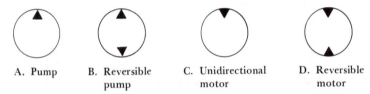

Figure 6–7   Schematic Diagrams of Rotational Components

As soon as the pressure overcomes the setting of the relief valve, the valve opens and returns the fluid to the tank.

Rotational Components

Circles are used in hydraulic drawings to illustrate rotational components. A **power triangle** is placed within the circle to denote energy sources (pumps) or energy receivers (motors). Figure 6–7 illustrates four types of rotational components.

In Figure 6–7A, the power triangle is pointing out from the center. Therefore, the component is an energy source, or a hydraulic pump. Figure 6–7B illustrates two power triangles pointing outward. This symbol means that the component is a pump and that it is also reversible. In Figure 6–7C, the power triangle is pointing into the center of the component. Therefore, the device receives energy and is a motor. Figure 6–7D illustrates the symbol for a reversible hydraulic motor.

HYDRAULIC ACTUATORS

A device that converts hydraulic energy into mechanical energy is called a **hydraulic actuator**. There are two categories of actuators:

cylinders and motors. Each category is discussed in the following subsections.

Cylinders

Cylinders are classified as linear actuators. The term *linear* is used because of the linear force developed by the output of the cylinder.

In hydraulic circuits, four types of cylinders are employed: ram cylinders, telescoping cylinders, double-acting cylinders, and double-rod cylinders. The following paragraphs describe these types in detail.

Ram Cylinder. The **ram cylinder** is illustrated in Figure 6–8. Its schematic diagram is shown in Figure 6–8A. With this cylinder, only one input of fluid may be directed to the fluid chamber. The ram cylinder provides a force causing the upward movement of the cylinder's rod. Thus, it has only one direction of movement: upward. The load applied to the head of the ram causes the ram to retract when pressure is removed.

Figure 6–8B illustrates the action of fluid flow into and out of the cylinder. Notice that when the fluid is flowing into the cylinder, the load is moved upward. The fluid is displacing the load and causing the load to rise. When the fluid is removed from the cylinder, the load is lowered.

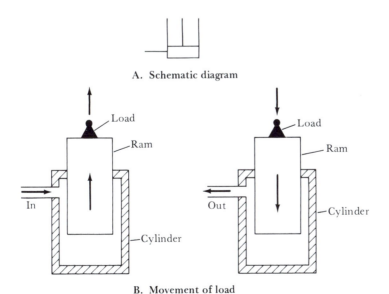

A. Schematic diagram

B. Movement of load

Figure 6–8   Single-Acting Ram Cylinder

Telescoping Cylinder. Figure 6–9 illustrates the telescoping cylinder. Like the ram, the **telescoping cylinder** has only one inlet for hydraulic fluid. The force developed by the fluid filling the fluid chamber causes the ram to extend upward (Figure 6–9A). Different heights or lengths are achieved when the various telescoping parts extend. Gravity causes the retracting of the telescoping parts (Figure 6–9B). The rate at which the telescoping rods return is controlled by limiting the amount of fluid leaving the cylinder.

Double-Acting Cylinder. The **double-acting cylinder** is shown in Figure 6–10. Figure 6–10A shows its schematic symbol. The double-acting cylinder has two oil ports, as shown in Figure 6–10B. These ports allow fluid to enter and exit in two directions. Thus, the cylinder can move in two directions.

In Figure 6–10B on the left, the fluid chamber is being filled from the top. This port is called the input port. The output port at the bottom of the cylinder handles the fluid flowing out of the cylinder. This fluid flow from top to bottom causes the load to be lowered. During the reverse cycle, shown on the right in Figure 6–10B, the fluid reverses direction. Fluid now fills the bottom of the chamber. The force generated by the fluid flowing into the bottom of the chamber causes the cylinder rod to rise and lift the load.

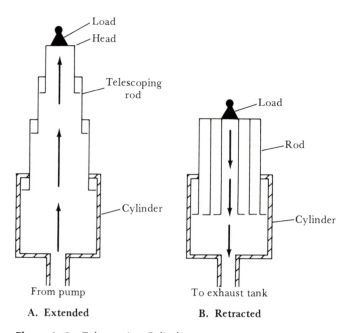

A. Extended   B. Retracted

Figure 6–9 Telescoping Cylinder

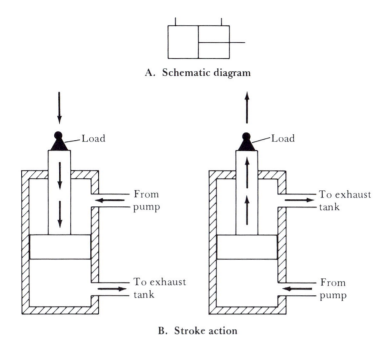

A. Schematic diagram

B. Stroke action

Figure 6–10 Double-Acting Cylinder

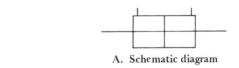

A. Schematic diagram

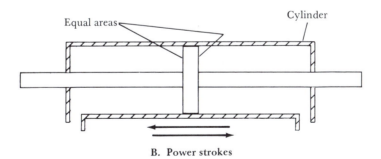

B. Power strokes

Figure 6–11 Double-Rod Cylinder

Double-Rod Cylinder. The **double-rod cylinder** is illustrated in Figure 6–11. Its schematic diagram is shown in Figure 6–11A. In Figure 6–11B, notice that the double-rod cylinder has the same number of ports as the double-acting cylinder. The major difference between the two types is that the double-rod cylinder has two rods, which are placed back to back. This design allows the cylinder to develop equal forces in both directions, along with equal speeds. With the double-rod leads from each cylinder, double loads can be connected to this device.

Accessories. Cylinder accessories include piston ring seals and cushions. The piston ring seals keep the oil within the cylinder and allow for the rapid cycling operation of the cylinder. Cushions provide a smooth operation to the cylinder when it comes to rest with large bearing loads applied to the end of the rod.

Motors

The second type of hydraulic actuator is the **hydraulic motor**, which is given its name because it is a rotational hydraulic actuator. The hydraulic motor resembles the hydraulic pump. The major difference is that a pump applies pressure to the hydraulic fluid, while a motor receives pressure from the fluid. This pressure causes the rotation of the motor. The rotating motor then develops a torque, providing a continuous rotational motion.

Figure 6–12 illustrates a typical hydraulic motor. The motor has two ports: the inlet port and the outlet port. The direction of the flow of the fluid determines the direction in which the motor will turn. In Figure 6–12, the fluid is entering the motor from the right-hand side, which causes the motor to turn in a counterclockwise direction. The outlet fluid leaves through the outlet port. If the positions of the inlet port and the outlet port were changed, the motor would then rotate in a clockwise direction.

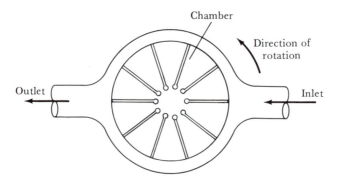

Figure 6–12 Hydraulic Motor

The following subsections discuss the ratings of hydraulic motors and the various types of motors used in robotic systems.

Ratings.   Hydraulic motors are rated by their displacement and their torque. **Displacement** is defined as the amount of fluid that the motor needs in order to turn one revolution. The motor displacement is given in cubic inches per revolution (in³/rev). For example, a typical hydraulic motor can develop a displacement of 10 cubic inches per revolution.

Torque is defined as the amount of force developed by the rotation of the motor. Generally, torque is the twisting effort of the motor. The torque of the motor must be large enough to overcome the friction and the resistance developed by the load. Torque is measured in pounds per inch (lb/in).

Figure 6–13 illustrates two general examples of torque measurement. In Figure 6–13A, the large pulley with a radius of 5 inches and a load of 5 pounds develops a torque of 25 pounds per inch at the center. In Figure 6–13B, the small wheel with a radius of 3 inches and the same load of 5 pounds develops a torque of only 15 pounds per inch at the center. Therefore, the radius of the pulley (or the motor) determines the torque developed by the pulley (or motor).

Types of Motors.   Two types of motors are found in hydraulic circuits: gear and vane. Figure 6–14 illustrates a **gear motor**. Torque in the gear motor is developed through the pressure of the fluid that is applied to the surface of the gear teeth. The gear teeth of the motor mesh and rotate together. The drive shaft to which the

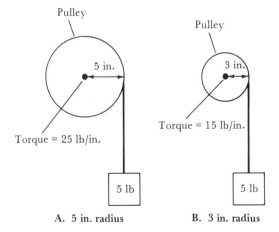

A. 5 in. radius   B. 3 in. radius

Figure 6–13   Torque Developed at the Center of a Pulley

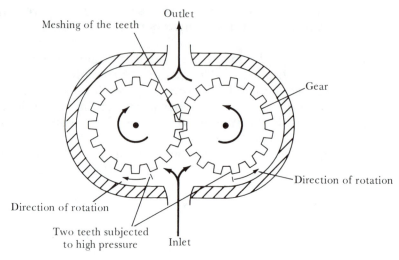

Figure 6–14   Gear Motor

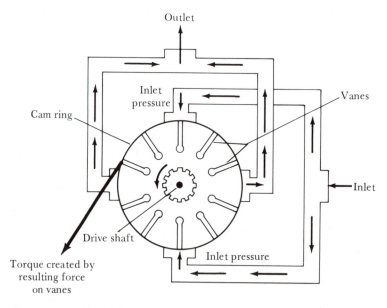

Figure 6–15   Vane Motor

load is applied is connected to only one of the gears. The motor can rotate in either direction by reversal of the flow of fluid through the motor. Gear motors operate in the range of 2400 revolutions per minute (rpm) and at a pressure of 2000 pounds per square inch.

Figure 6–15 illustrates a typical **vane motor**. Torque is developed by inlet fluid pressure on the exposed surfaces of the rectangular vanes. These vanes slide in and out of slots in the rotor's spindle. The spindle is connected to the drive shaft of the motor, and the shaft is connected to the load. Fluid from the two inlet ports flows into the top half and the bottom half of the fluid chambers between the vanes. The vanes seal the fluid between the cam ring and the spindle. The fluid flow causes the shaft to rotate. As the rotating vanes come to an outlet port, the fluid is returned to the reservoir for another cycle of operation.

In robotics, a hydraulic motor is usually selected because of its ability to accelerate at high speeds and to develop the proper amount of torque and fluid flow through the motor.

DIRECTIONAL CONTROLS

Directional controls, or **valves**, allow the hydraulic fluid to be directed through different paths. These paths control the operation of the motors or cylinders. Most of the valves used in the control of fluids have **finite positioning**. This term means that the valves control the paths that the oil takes by opening and closing flow paths through the valve. The hydraulic symbol for flow control valves is a square box containing an arrow. This arrow represents the flow path through the valve. The basic valves used in the operation of hydraulic circuits are check valves, globe valves, gate valves, spool valves, and multipath valves.

A **check valve** can operate either as a directional control or as a pressure control. Typical pressures for the check valve are well below 3000 pounds per square inch. In its simplest form, the check valve only allows passage of fluid in one direction; it blocks all flow of fluid in the opposite direction. Figure 6–16A illustrates fluid flow in the check valve. Notice that the fluid flows from the left side to the right side of the valve. In this case, the fluid flows freely through the valve. When fluid passes from the right side to the left side, the ball is forced against the seat, as shown in Figure 6–16B. In this case, the flow of fluid through the valve is stopped. Many times, a spring is placed behind the ball on the right side, as indicated in Figure 6–16A. This spring ensures that the ball will have a proper seat.

Figure 6–16C illustrates two graphic symbols that can be used for the check valve. The composite valve symbol illustrates either an open valve or a closed valve. This graphic diagram is very complex for a simple check valve. The graphic symbol for the simple

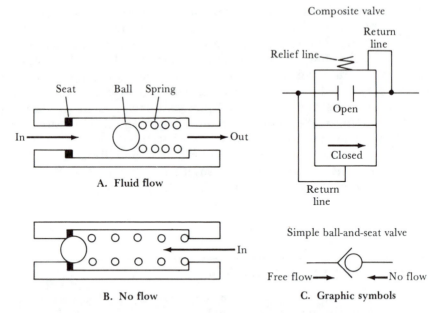

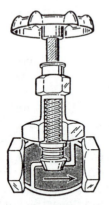

A. Fluid flow

B. No flow

Composite valve

Simple ball-and-seat valve

C. Graphic symbols

Figure 6–16 Check Valve

Figure 6–17 Globe Valve

ball-and-seat valve is the symbol most often used in hydraulic schematic diagrams.

Figure 6–17 illustrates the design of the **globe valve**. It can withstand pressures up to 150 pounds per square inch. This type of valve works well when it is fully open or fully closed.

Figure 6–18 illustrates the **gate valve**. This valve is used when the application requires higher pressures and greater flow than can be achieved with the globe valve. A typical maximum pressure for

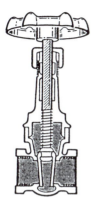

Figure 6–18   Gate Valve

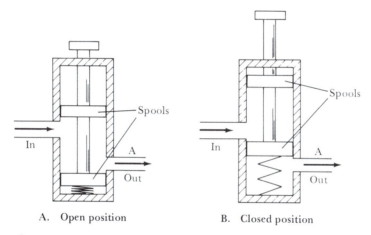

A.   Open position   B.   Closed position

Figure 6–19   Spool Valve

this valve is 5000 pounds per square inch. Like the globe valve, this valve will operate fully open or fully closed.

The **spool valve**, or **piston valve**, is the most common valve used in flow control. It can withstand pressures up to 3000 pounds per square inch. Figure 6–19 illustrates the spool valve. The name for the valve is derived from the fact that the components on the internal part of the valve look like spools. When the piston is down, as in Figure 6–19A, the valve is open. When the piston is raised, as in Figure 6–19B, the valve is closed. The spool valve may be controlled by several methods, such as manual operation, electric solenoid operation, or operation by another hydraulic system.

Multipath control valves provide one inlet port and several output ports. For example, Figure 6–19 illustrates a two-way spool valve. When fluid enters the input port, it is directed to path A. When fluid enters by way of path A, it is directed to the input port.

HYDRAULIC PUMPS

The most important section of the hydraulic drive of a robotic system is the **hydraulic pump**. The pump is used in the hydraulic circuit to convert mechanical energy into hydraulic energy. The pump accomplishes this conversion by pushing the hydraulic fluid through the system. Pumps are classified into two types: hydrodynamic pumps and hydrostatic pumps.

The **hydrodynamic pump** employs water as the fluid under pressure. It is used for fluid transfer when the fluid will encounter low resistance. The pump consists of an inlet and an outlet. Fluids entering through the inlet are transferred to the outlet by a turning impeller. These pumps do not create much pressure and hence are not generally found in robotic hydraulic drives.

Hydrostatic pumps are the type usually found in robotic hydraulic drives. They use hydraulic fluid as the fluid under pressure. These pumps provide a large amount of output fluid for every stroke.

The following subsections discuss the ratings of hydraulic pumps and the types of pumps used in robotic systems.

Ratings

A pump is rated by its maximum operating pressure and its output, in gallons per minute, at a given drive speed. A second rating of a pump is its pressure rating, which is specified by the manufacturer. Pressure rating tells the user the overall length of service life of the pump. If a pump is operated at higher pressures than those specified by the manufacturer, reduced pump life or more serious damage may result. The third rating of a pump is its displacement. Displacement determines the output of fluid from the pump, and it is given in gallons per minute or cubic inches per revolution.

Types of Pumps

Two types of pumps are found in hydraulic circuits: gear and vane. These pumps must deliver constant flow of fluid to the various axes of the manipulator. The fluid delivered allows movement of an axis and creates the lifting power of the manipulator.

Figure 6–20 illustrates the **gear pump**. Notice that this pump looks very much like the hydraulic gear motor. A partial vacuum

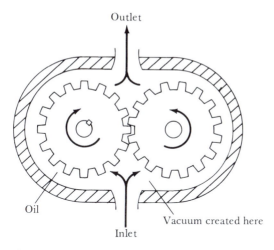

Figure 6–20 Gear Pump

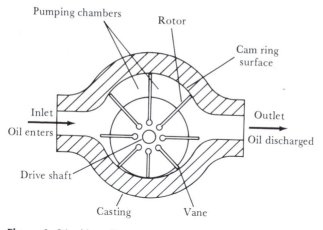

Figure 6–21 Vane Pump

is created at the inlet as the geared teeth unmesh. Fluid flows into the pump to fill the space and is carried around the outside of the gears. As the teeth mesh again at the outlet, the fluid is forced out.

Figure 6–21 illustrates the **vane pump**. Fluid flowing into the pump from the inlet side is trapped in the pumping chambers created by the vanes of the pump. The fluid is pressed against the vanes of the pump, which fit into a slotted rotor. Centrifugal force pushes the vane to the outside of the cam, where the fluid is trapped. As the fluid presses against the vanes, the force causes the pump to rotate around the rotor. Finally, the fluid exits at the outlet of the pump. High pressure at the pump's outlet imposes an unbalanced load on the gears and the bearings supporting them.

The vane pump can deliver high speed and high pressure to the actuator. This combination enables the robot manipulator to move, lift, or perform various tasks. The vane pump is capable of lifting the fluid in the hydraulic system to a height of 25 feet. Generally, in robotic systems, this pump will provide a pressure as high as 2000 pounds per square inch and will develop a discharge capacity of as much as 35,000 gallons per minute.

HYDRAULIC RESERVOIRS

The operation of the hydraulic system depends on fluids. However, the fluids required for the operation of the robotic manipulator cannot be stored in the hydraulic lines. Instead, the hydraulic fluids are stored in a **reservoir**. The fluid in the reservoir is kept clean by the use of filters.

Figure 6–22 illustrates a typical reservoir for hydraulic operations. Notice that the reservoir contains several components: the inlet line, the baffle plate, the filter, the drain plug, the return line to the pump, and the reservoir housing.

The ends of the inlet and outlet lines in the reservoir must be lower than the fluid level. If they are not, the fluid will become aerated and foam, a condition called **turbulence**. Turbulence in the reservoir decreases the hydraulic system's efficiency. That is, tur-

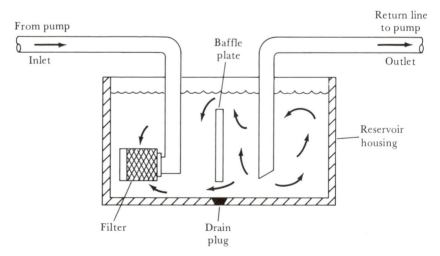

Figure 6–22   Reservoir

bulence in the reservoir creates air in the fluid, which, in turn, causes the temperature of the fluid to increase.

The **baffle plate** is used to separate the inlet side from the outlet side of the reservoir. Generally, the baffle plate is about two-thirds of the height of the oil level in the reservoir. The baffle plate has the following four functions:

1. It prevents local turbulence in the tank.
2. It removes trapped air from the fluid.
3. It separates foreign material from the fluid and allows this material to settle to the bottom of the tank.
4. It increases the heat transfer through the reservoir walls.

The **filter** in the reservoir is used to trap all foreign materials in the fluid. The drain plug is used to drain the fluid from the tank so that it can be cleaned or so that routine maintenance can be performed.

PNEUMATIC SYSTEMS

The **pneumatic system** for robotic manipulators is very similar to the hydraulic system just discussed. The major difference is that the hydraulic system transfers fluid under pressure, while the pneumatic system transfers air under pressure.

Both the hydraulic system and the pneumatic system are used for accurate positioning and accurate repeatability of the robot's manipulator. Hydraulic systems are generally used when heavy weights must be positioned. Pneumatic systems are generally used when light loads are to be positioned. In many applications, the pneumatic and hydraulic systems operate together. For example, the hydraulic system may be used to lift a heavy part, and the pneumatic system may be used to clamp the part into position. Also, a small pneumatic system may be used to control the hydraulic valves in the system.

Other differences between the hydraulic system and the pneumatic system involve overall costs and components. The hydraulic system is costly to operate and requires special components. Also, the fluids in the hydraulic system must be stored in a reservoir so that they can be reused in the system. In contrast, the pneumatic system uses atmospheric air; when the air has finished its job, it is returned to the atmosphere. Thus, the pneumatic system is not costly to operate.

The basic components in the pneumatic system are illustrated in Figure 6–23. The system contains an air compressor and an air

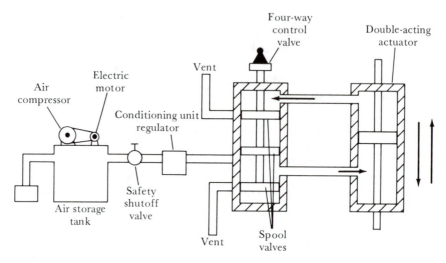

Figure 6–23   Basic Components of a Pneumatic System

storage tank. These components are used to build up the pressure in the system. A safety shutoff valve is placed between the generating source (the electric motor) and the load. The drive system for the pneumatic system is the electric motor. The electric motor connected to the air compressor maintains pressure in the system.

The conditioner in the air line supplies the air with a slight amount of oil. The oil reduces the friction that can be developed as the air passes through the cylinders or the air lines. The regulator keeps the pressure in the line regulated. The vent is much like the relief valve in the hydraulic supply. When the pressure builds up beyond the rated value of the system, the vent allows the excess pressure to escape.

The four-way control valve allows the system to operate either the double-acting actuator or the spool valve to lift the load. The spool valve controls the direction of airflow. It causes the air to flow in a certain direction and therefore makes the actuator move up or down. The double-acting cylinder moves the load up or down, depending on the air pressure in the line. The control valves and the actuators in the pneumatic system are very similar to their counterparts in the hydraulic system, which were described earlier in the chapter.

SUMMARY

The hydraulic drive supplies the power needed for the movement of the robotic manipulator. The hydraulic system uses fluids (usually oil) under pressure. The pressure developed drives the various hydraulic components and converts hydraulic energy into mechanical energy for manipulator movement.

Pressure is defined as the amount of force applied to the surface area of a fluid when the fluid is forced to flow. Pressure is measured in pounds per square inch. The fluid under pressure can be water or hydraulic fluid. When water is used, the pump is called a hydrodynamic pump. When hydraulic fluid is used, the pump is called a hydrostatic pump.

Pressure in the hydraulic system is directly related to the distance the fluid must travel. The pressure is greater at the inlet end of the pipe. As the fluid flows away from this point, the pressure is reduced.

The velocity of a fluid is the average speed at which the fluid particles pass a given point during a given period of time. The velocity of a fluid is measured in either feet per minute, feet per second, or inches per second. Flow rate is the measurement of the volume of fluid passing a given point in a given time period. Flow rate is measured in gallons per minute or cubic inches per minute.

Hydraulic fluid has two types of energy: kinetic energy and potential energy. Kinetic energy is developed from the weight of the fluid. Potential energy is developed from the pressure of the fluid. The potential energy and the kinetic energy in the hydraulic system must be constant at all points in order for the pump to develop a constant flow rate. This law is known as Bernoulli's principle.

Graphic symbols are used in schematic drawings to represent the various hydraulic components. The pipes or lines that connect the components in the hydraulic circuit are drawn as straight lines. There are three types of lines: pressure lines, pilot lines, and drain lines. The rotational components are either hydraulic motors or pumps. Pumps are used to output energy, while motors are used to input energy. Their graphic symbols are circles with power triangles inside. Triangles pointing outward indicate output energy, while triangles pointing inward indicate input energy.

Hydraulic actuators are devices—cylinders or motors—that convert hydraulic energy into mechanical energy. There are four types of hydraulic cylinders used with manipulators: the ram cylinder, the telescoping cylinder, the double-acting cylinder, and the double-rod cylinder. Hydraulic motors are rotational hydraulic actuators. Motors are rated by the amount of torque developed and by the amount of displacement. Two types of motors are employed in robotics: the gear motor and the vane motor.

Directional control of the hydraulic fluid is very important for the operation of the hydraulic system. The simplest form of directional control is the check valve. The globe valve and the gate valve are used to stop all flow of fluid in a main trunk of the hydraulic system. Spool valves or piston valves are used for positioning the load.

The hydraulic pump is used to convert mechanical energy into hydraulic energy. There are two types of pumps that are employed in hydraulic systems: the hydrodynamic pump and the hydrostatic pump. The hydrodynamic pump is used in low-resistance systems. The hydrostatic pump is used in drive systems requiring a large amount of output fluid.

The device used to store hydraulic fluids is the reservoir. The reservoir contains an inlet line, a baffle plate, a filter, a drain plug, and an outlet line.

Many manipulators are driven by pneumatic systems. The pneumatic system operates much like the hydraulic system, but it uses air under pressure rather than fluid under pressure. The pneumatic system has an air storage tank, a compressor, safety valves, regulators for directional control, and actuators for movement of the load.

The hydraulic and the pneumatic systems may operate together to form a total working cell. The hydraulic system may be used to lift heavy loads, and the pneumatic system may be used to position lightweight parts.

KEY TERMS

baffle plate	hydraulic flow
Bernoulli's principle	hydraulic motor
check valve	hydraulic pump
cylinder	hydraulics
directional controls	hydrodynamic energy
displacement	hydrodynamic pump
double-acting cylinder	hydrostatic energy
double-rod cylinder	hydrostatic pump
drain line	kinetic energy
filter	multipath control valve
finite positioning	pilot line
flow rate of a fluid	piston valve
gate valve	pneumatic system
gear motor	potential energy
gear pump	power triangle
globe valve	pressure
hydraulic actuator	pressure drop

pressure line
ram cylinder
reservoir
spool valve
telescoping cylinder
torque

turbulence
valves
vane motor
vane pump
velocity of a fluid

QUESTIONS

1. When the movement of a manipulator's arm uses pressurized fluid, what is the energy called?

2. What is the value of atmospheric pressure at sea level?

3. Which component is used to create the flow of fluid in a hydraulic system?

4. Define the term *velocity of a fluid*.

5. What units are used for the measurement of velocity of a fluid?

6. What condition exists when fluid flows?

7. What factor develops pressure as fluid flows through a pipe?

8. At which point is the pressure the greatest in a hydraulic system?

9. What principle states that the rate of fluid flow remains constant as the cross-sectional area of the pipe changes over a given length?

10. Which line in a hydraulic system carries the majority of fluid to various devices in the hydraulic circuit?

11. For which component in a hydraulic schematic diagram does the power triangle point away from the center?

12. Name the hydraulic component that is used to convert hydraulic energy into the mechanical energy of the manipulator.

13. What type of hydraulic actuator can move in two directions?

14. What term identifies the amount of fluid required to turn a hydraulic motor one revolution?

15. The hydraulic motor must develop a twisting action to overcome the resistance of the load. What is the term used for this twisting action?

16. Which section of the motor is connected to the load?

17. Which component in a hydraulic circuit allows the flow of a fluid in one direction?

18. What is the maximum amount of pressure that can be applied to a gate valve?

19. Which component develops a constant flow of fluid to the various axes?

20. Describe the ratings of a hydraulic pump.

21. Draw the schematic symbol for a hydraulic reservoir.

22. Why does the baffle plate in the reservoir separate the input line and the output line?

23. Using a hydraulic schematic, identify the various components of the hydraulic system.

24. Why is a conditioning unit needed for a pneumatic system?

7

DC and AC Motor Operation

OBJECTIVES

Upon completing this chapter, you should be familiar with:

— DC motors,
— Speed control of the DC motor,
— Reversing a DC motor,
— Stopping the motor,
— Stepper motors,
— DC brushless motors,
— AC induction motors.

INTRODUCTION

As we have discussed, the manipulator's drive power can be developed from three basic sources: hydraulic power, pneumatic power, and electric power. Hydraulic power allows the robot manipulator to lift heavy loads. Pneumatic power allows the manipulator to lift and position light payloads. Electric power allows the axes to lift and position medium to heavy loads. Chapter 6 described hydraulic and pneumatic drives; this chapter examines the electric drive system.

Electric drives are of two types: the DC motor and the AC motor. These motors, when used as drive mechanisms, are equipped with positional information systems that continually feed positional information back to the controller. This feedback signal is called the *tachometer signal*. It is generated from a simple DC generator.

For all robotic manipulators, the velocity of the axes as they approach a programmed point in the work cell is very important. So that the velocity of the axes remains constant, the motors used for manipulators have speed controls. The speed of the motor is controlled through a circuit that regulates the current flow to the shunt field of the motor or to the armature field of the motor.

Once the manipulator has been placed into position, it must remain in that position—that is, the motor must stop. One method of stopping the motor is dynamic braking. A second method discussed in this chapter is by a plugging action on the motor.

A stepper motor is generally used with manipulators in the medium-technology area. Stepper motors must have electronic pulses applied to their windings to give them movement. Each pulse applied to the motor allows the motor to rotate a certain number of degrees.

Many high-technology manipulators are driven by DC brushless motors or by AC motors. These motors require no brushes and thus do not generate sparks in their operation.

The major portion of this chapter is devoted to a discussion of DC motors: their characteristics, speed control, braking methods, and types. The final section of the chapter examines AC induction motors.

DC MOTORS

An **electric motor** is a device that converts electric energy into mechanical energy. The motor accomplishes this conversion by using DC power or AC power. Through this conversion process, the manipulator's axes can be moved to the desired locations.

The DC motor is used in drive systems for two reasons. The first reason involves the relationship between the speed and the torque of the DC motor. The torque of the DC motor can change over a wide range of applications. That is, as the load applied to the motor increases, the torque of the motor also increases. But this increased torque tends to slow the motor down. Additional current supplied to the motor will overcome the torque and keep the speed of the motor constant. The second reason DC motors are used is that DC motors can easily be interfaced with electronic components. Thus, the DC motor can be controlled by microprocessors and by other electronically controlled 5-volt DC logic.

In this section, we will examine the principles of operation of DC motors and the common types of motors used in robotic drive systems.

Principles of Operation

The **DC motor** operates on the principle of two interacting *magnetic fields*. One of the magnetic fields is created by permanent magnets, and the second field is created by electromagnetics.

Figure 7–1 illustrates the basic components of a simple DC

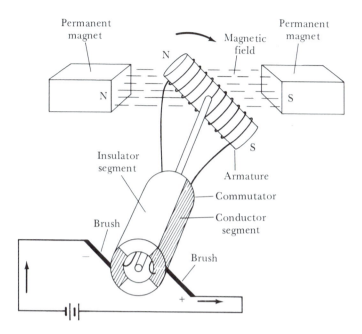

Figure 7–1   Basic Components of a DC Motor

motor. The **permanent magnets** form one magnetic field. A wire-wound **armature** creates the second magnetic field. The armature of the DC motor is allowed to rotate. Therefore, the armature can also be called a **rotor**.

Another component of the DC motor is the commutator. The **commutator** acts as a double-pole, double-throw switch. Part of the commutator is a conductor, while the other part of the commutator is an insulator. These two parts allow the rotor coil to develop reversible magnetic fields. Each end of the rotor coil is connected to one of the commutator's conducting parts.

The electromagnetic field is developed from an outside power source, that is, from the connection of a DC power supply to the motor. This connection is made to the armature winding through components called **brushes**. The brushes touch the commutator's conductor. As the rotor turns, the brushes pass against the conductive and insulating parts of the commutator.

Figure 7–2 illustrates the rotational action of the DC motor. In Figure 7–2A, a DC current is supplied to the brushes of the motor. The brushes of the motor pass the DC current to the commutator. Notice that the commutator has two conductor segments, labeled 1 and 2 in the figure. These segments allow the DC power supply to complete its path through the armature windings. The

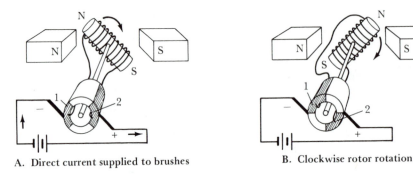

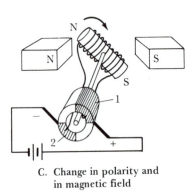

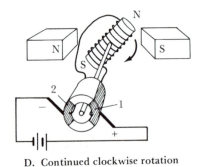

Figure 7–2 Rotating Action of a DC Motor

current is then passed to the armature and to the armature windings. A magnetic field is thus built up around the armature windings. This magnetic field has the same polarity as the field of the permanent magnets. Because both magnetic fields have the same polarity, the two fields repel each other. The armature thus starts to move in a clockwise direction.

The magnetic field causes the armature to rotate 90°. The polarity of north/south established on the armature remains on the armature. But as the armature turns, the magnetic north pole of the armature comes under the pull of the south pole of the permanent magnet, as illustrated in Figure 7–2B. Since opposite poles attract, the armature continues to rotate an additional 90°.

This rotation of the commutator changes the polarity of the DC power passed to the armature windings. The change in polarity causes the magnetic field in the windings of the armature to change. Figure 7–2C illustrates this action. Notice that the positive part of the supply is now connected to segment 1 of the commutator. Thus, the end of the armature that was north in Figure 7–2A is now south.

The armature's magnetic field and the permanent magnetic field now have the same polarity, and they repel each other. The commutator continues to supply this polarity for another 90° of rotation in the clockwise direction. But, again, as the armature turns, the south pole of the armature comes under the influence of the north pole of the permanent magnet. This action continues the armature's rotation in the clockwise direction, as shown in Figure 7–2D.

The rotation just described continues during the entire time that DC power is supplied to the armature. The action of the two magnetic fields operating together causes the rotational movement of the DC motor.

Motor Types

The DC motors used for the manipulator's axes have three basic designs: series-wound design, shunt-wound design, and compound-wound design. Each of these designs relates to the type of winding used on the field coils of the motor in reference to the armature windings. The **field coils**, or field windings, of the motor provide the magnetic field in which the armature of the motor rotates. Generally, the field coils are connected to a DC power supply, which provides the current necessary for the development of the magnetic field. Each of the three types of DC motors is described in the following subsections.

Series-Wound Motor.   A schematic diagram of the **series-wound motor** is shown in Figure 7–3. In this motor, the armature windings are connected in series with the armature of the motor. The field windings have a very low resistance. That is, only a few turns of large wire are used in this winding. The low resistance causes a high current to flow in the circuit when the motor starts. The high current, in turn, develops a high torque in the motor.

In Figure 7–3, the applied voltage V_{AC} is connected to the motor circuit. This power supply provides the current through the series

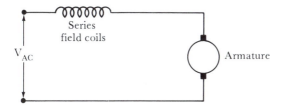

Figure 7–3   Schematic Diagram of a Series-Wound DC Motor

field coils that is necessary to develop the magnetic field. Once the magnetic field is developed, the motor will begin to rotate.

The speed of the series-wound motor depends on the current through the windings. That is, the current through the windings causes the field strength in the motor to vary. The increase or decrease in field strength causes the speed of the motor to vary.

The current through the windings also changes when the loads connected to the motor change. When the loads are light, the current through the motor is low. Therefore, the speed of the motor is high. As the load increases, the speed of the motor is reduced. Thus, the series-wound motor must have a load connected to it. If no load is connected, this motor will rapidly accelerate, causing the motor to be damaged beyond repair. The rapid acceleration of the motor is called a **runaway condition**.

In robotics applications, the series-wound motor is often used for the wrist drive. The high speed of this motor allows the wrist of the manipulator to reach the high speeds needed for quick positioning of the end effector.

Shunt-Wound Motor.   Figure 7–4 presents a schematic diagram of the **shunt-wound motor**. Notice that the shunt field coils are connected in parallel with the armature of the motor. The field coils have a very large resistance. This large resistance is developed from many turns of small-gage wire wound in parallel with the armature circuit. The field strength of this circuit can be changed by the use of the variable resistor that is connected in series with the field windings. The change in field strength causes the motor's speed to vary.

The parallel connection of the field coils provides another benefit to the shunt-wound motor. The parallel connection develops a constant voltage across the motor. This constant voltage maintains a constant motor speed. The constant motor speed means that the motor develops constant torque.

Under changing load conditions but with constant applied voltage, V_{AC}, the shunt-wound motor can develop variable torque with

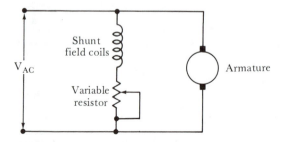

Figure 7–4   Schematic Diagram of a Shunt-Wound DC Motor

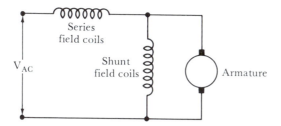

Figure 7–5 Schematic Diagram of a Compound-Wound DC Motor

constant speed. Also, the shunt-wound motor can develop adjustable speeds, since the DC supply to the motor can be varied. Hence, the shunt-wound motor is more flexible than the series-wound motor.

Compound-Wound Motor. The **compound-wound motor** has two sets of windings. One set is a series-wound coil, and the other set is a shunt-wound coil. Figure 7–5 shows a schematic diagram of the compound-wound motor.

   The compound-wound motor has a set of coils connected in parallel with the armature. The speeds at which this motor operates are generally specified as full-load speed and no-load speed. The connection of the coils in series and in parallel gives this motor the same operating characteristics as the series-wound and shunt-wound motors.

SPEED CONTROL OF THE DC MOTOR

The speed of a DC motor can be controlled through regulation of the windings' voltage and current or through electronic circuit control. Two types of motor control are used in DC motor operation: shunt field control and armature control.

   Shunt field control adjusts the voltage (or current) applied to the field windings. As the field voltage is increased, the motor's speed will decrease. Armature control adjusts the voltage (or current) applied to the armature. As the voltage on the armature is increased, the motor's speed will increase. Armature control can be accomplished through electric circuit devices or through solid-state, electronic circuit devices. Each type of control is described in the following subsections.

Shunt Field Control

The schematic diagram for **shunt field control** is shown in Figure 7–6. Here, a variable resistor R is connected in series with the field

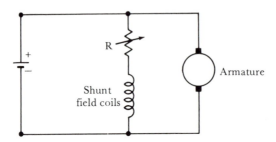

Figure 7–6   Shunt Field Control

windings of a shunt-wound motor. This variable resistor controls the amount of voltage developed across the field windings. The field voltage is increased by reducing the value of variable resistor R. The result is a stronger magnetic field, which induces a large counter electromotive force (EMF) in the armature windings. This large counter EMF opposes the DC voltage applied to the armature and reduces the armature current of the motor. The reduction of armature current then reduces the speed of the motor. The decrease in motor speed will continue until the counter EMF returns to normal. Therefore, a decrease in armature current slows the motor's rotational capabilities.

The opposite reaction increases the speed of the motor. When the resistance value of R is increased, the field current in the motor is reduced. The reduced field current decreases the magnetic field in the field windings. This reduction of the magnetic field allows more current to pass through the armature, thus making the armature spin faster. The counter EMF produced by the field windings is small and therefore develops no opposition to the armature current. So, the motor speed increases.

Shunt field control of motor speed is very useful in low-cost projects. That is, an inexpensive variable resistor can be used to control the windings' current and hence to control the speed of the motor.

Shunt field control has one drawback regarding the torque developed by the motor. Since the torque of the motor depends on the current of the armature and the strength of the magnetic field, this motor control reduces the torque that the motor can develop at high speeds. And since the load requires torque from the motor, this motor control process reduces the flexibility of the system.

Armature Control

Armature control allows the motor to develop the necessary torque at high speeds. In this control system, the armature voltage and

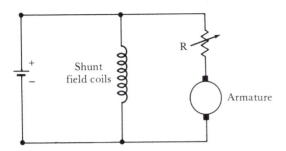

Figure 7–7   Armature Control

current are increased by a reduction of the value of resistance. In Figure 7–7, notice that the variable resistor is connected in series with the armature of the motor. When power is initially supplied to this system, the motor starts to run faster, producing greater amounts of torque. The increase in speed results because the armature is developing greater amounts of voltage. But the increase in voltage to the armature—and the simultaneous faster spinning of the armature—causes an increased counter EMF to be developed. The counter EMF then reduces the armature current. In turn, the motor speed is reduced.

The one drawback of this system is that the variable resistor in series with the armature must be able to handle the armature current. Since the current feeding the armature is large, the variable resistance must also be large. Thus, the variable resistor must be very large in physical size, and a large resistor increases the cost of the motor control system. Of the two types of motor control systems, shunt field control is preferable to armature control, because of its control of current to the armature.

Solid-State Control

Another method for controlling the speed of the motor is **solid-state armature control**, accomplished through the use of solid-state components. Generally, these components are silicon-controlled rectifiers (SCRs).

Figure 7–8 shows a simple electronic control circuit that might be used for motor speed control. An AC source is supplied to the circuit for operational power. The AC 60-hertz voltage goes to a full-wave bridge rectifier circuit with four diodes. The rectifier circuit converts the AC voltage into a DC operating voltage.

The DC output of the bridge rectifier is applied directly to the field windings of the motor. The bridge's output voltage is also applied to the armature windings of the motor. The low resistance of the armature windings allows the capacitor C to charge through diode D_2. The variable resistor R_2 controls the rate at which the capacitor C will charge. The charging rate of C continues until C is charged to the break-over voltage of the four-layer diode.

The break-over voltage of the four-layer diode is the gate voltage of the SCR. When the gate voltage of the SCR is reached, the SCR will conduct. The conducting SCR develops a path to ground for the armature current and thus reduces the speed of the motor.

The SCR provides half-wave rectification and control for the armature windings. The gate control circuit controls the firing time of the SCR. When the SCR is triggered early, the average current passing through the SCR is large. This large current flow increases the current supplied to the armature, thus causing the motor to run faster. When the firing of the SCR is delayed, the average current supplied to the armature is smaller. The smaller current causes the motor to run slower.

Diode D_3 is placed in the circuit to provide a path for any inductive kick developed in the motor. This inductive kick will produce a counter EMF in the motor that will damage the windings of the motor. Diode D_1 and resistor R_1 are used to provide a discharge path for the capacitor C so that it can recharge. The SCR in this motor control circuit only turns on for one-half cycle of operation of the AC input waveform.

The solid-state circuit shown in Figure 7–8 is only one type of motor speed control. Other circuit configurations can be employed for better speed control.

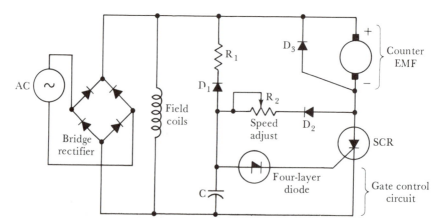

Figure 7–8 Solid-State Control

═══ REVERSING A DC MOTOR

Many times in the operation of the robot's manipulator, the DC motors must be able to rotate in two directions. For instance, the reach axis of a cylindrical robot must reach out and pick up the part and then return to its original position. This reach-and-return action requires a reversible DC motor operation.

Most DC motors can be **reversed** in rotational direction by a change in the polarity of the DC voltage applied to the armature of the motor. The switching action generally takes place in the armature, because the armature lowers inductance. The lower inductance causes less arcing of the switching contacts when the motor reverses its direction.

The switching operation should take place when the motor has come to a stop. So that the motor can handle the quick rotation changes, the motor's control circuits must be designed for the switching operation. The reversal of the motor can be accomplished in two ways:

1. Reversing the direction of the armature current and leaving the field current the same;
2. Reversing the direction of the field current and leaving the armature current the same.

Figure 7–9 illustrates two methods for reversing the direction of the DC motor: mechanical control (Figure 7–9A) and solid-state control (Figure 7–9B). In Figure 7–9A, a simple contactor directionalizes the DC motor. The forward (FOR) contactor causes current to flow through the armature in one direction, while the reverse (REV) contactor, when closed, causes current to flow through the armature in the opposite direction.

In Figure 7–9A, the FOR contactor is closed when the forward start button is pressed. Closing the FOR contactor causes the current flow through the armature to rotate the motor in a clockwise direction.

The contactor is used as a switch in this circuit to direct current flow through the armature of the motor. In the figure, the contactors shown with a line through them are normally closed contacts; the contactors with no line through them are normally open contacts. When the forward or reverse switch is closed, the open contactors are automatically closed and the closed contactors are automatically opened.

The motor will rotate in a clockwise direction until the stop button is depressed. When the stop button is depressed, the FOR contactor opens, thus stopping the motor.

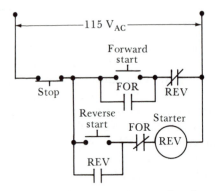

A. Mechanical control of motor direction

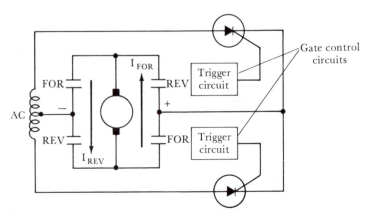

B. Solid-state control of motor direction

Figure 7–9 Motor-Reversing Circuits

When the reverse start button is depressed, the REV motor starter is energized. Now, the current will flow in the opposite direction. Hence, the motor will rotate in a counterclockwise direction.

Figure 7–9B is a simple schematic diagram of solid-state control of motor direction. Notice that SCRs are used in the circuit for motor speed control. The gate control circuits of the SCRs provide the speed control operation. The contacts are again used to control the rotation of the motor. When the FOR contacts are closed, the motor develops current flow in the I_{FOR} direction. When the REV contacts are closed, the motor develops current flow in the reverse, I_{REV}, direction.

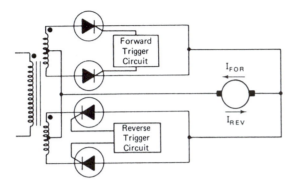

Figure 7–10   Motor-Reversing with Total Solid-State Circuitry

The angle at which the SCR will fire is controlled by the trigger circuits. The trigger circuits control the speed of the motor by controlling the angle at which the SCR will conduct.

Figure 7–10 illustrates the motor-reversing method employed for total solid-state circuits. The total reversing ability of this circuit depends on the circuits located in the forward and reverse trigger circuits. Notice that two sets of SCRs are provided. One set is used for current flow in one direction through the armature of the motor. This forward current flow, I_{FOR}, causes the motor to rotate in the clockwise direction. The forward current is controlled by a directional control circuit not shown in the diagram.

The second set of SCRs is used for current flow in the opposite direction. When the directional control circuit for the reverse direction is energized, the forward circuit is disabled. The reverse circuit causes the current to flow in the opposite direction. This reverse current is labeled I_{REV} in the figure.

STOPPING THE MOTOR

In many applications, a manipulator must be stopped and held in position. Hence, the DC motor must be able to be stopped and held in position, too. The action of stopping the motor rotation is called **braking**. Two methods can be employed when braking is needed: an electromechanical, or dynamic, braking method and a plugging method. Both methods are described in the following subsections.

Braking applications are very important in the movement of the axes of the robot's manipulator. When an axis is resting in a certain location, the braking circuits of the manipulator turn on automatically in order to keep the axis in the desired position. When the operator of the robot moves the axis, the braking circuits turn off, and the axis is free to move again.

Dynamic Braking

Figure 7–11 illustrates the circuit used for **dynamic braking**. Notice that the armature of the motor can be connected to the power supply line and also to the braking circuit. In the figure, the armature of the motor is connected to the terminals labeled position 1. This position is the normal operating mode for the motor. The dynamic-braking circuit is located at the terminals labeled position 2. In this circuit, a simple resistor R is connected across the terminals.

When braking is needed, the armature windings of the motor are mechanically connected across resistor R and the terminals labeled position 2. The voltage generated by the armature will now develop a reverse current flow through the armature. This reverse current flow produces a counter EMF and thus slows the rotation of the motor. Generally, dynamic-braking action incorporates some type of mechanical relay, which is used to switch R in and out of the circuit.

Plugging

The **plugging** method of braking is illustrated in Figure 7–12. In the figure, the armature of the motor is connected to the input circuitry (V_{DC}) through position 1 of the switch. This circuit connection allows the motor to rotate in one direction. When the switch is moved to position 2, the armature of the motor is connected with the opposite polarity across the input line voltage. This connection causes the current through the armature to flow in the reverse direction, which, in turn, slows the motor down. When the motor

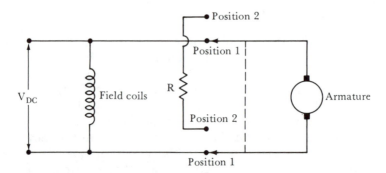

Figure 7–11 Dynamic Braking of a Motor

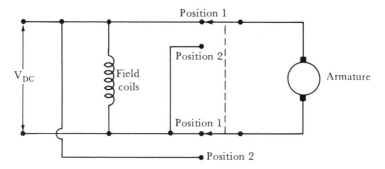

Figure 7–12   Plugging a Motor

comes to rest, the switch contactors disconnect the motor from the line and reconnect the motor for normal directional rotation.

STEPPER MOTORS

In many low-technology robotic systems, the stepper motor is used as the manipulator's drive mechanism. The **stepper motor** operates through pulses supplied to the motor. Each pulse causes the motor to rotate in a clockwise or a counterclockwise direction, and each pulse causes a certain degree of rotation. Therefore, several pulses applied to the motor supply a certain degree of movement. Thus, the stepper motor allows the axis of the robot to be placed into position.

In this section, we will discuss stepper motor action and the types of stepper motors used in robotics.

Stepper Motor Action

The stepper motor has the ability to move in a clockwise or a counterclockwise direction, to start or stop at various mechanical stops, and to move in precise angular increments for each input pulse supplied to the motor. As these pulses are inputted to the motor, the motor's armature will rotate a given number of degrees. The number of degrees that the motor rotates depends on motor size.

The stepper motor comes in a variety of sizes; each size relates to a specific rotation of the rotor. The following table lists the sizes of typical stepper motors used in robotics and the degrees of rotation developed by each motor:

Size of motor	Degrees of rotation per pulse
240	1.5°
180	2.0°
144	2.5°
72	5.0°
48	7.5°
24	15.0°
12	30.0°

Notice that a stepper motor of size 240 rotates 1.5° for each pulse input. A stepper motor of size 12 rotates 30° for each pulse input. Each stepper motor is built for a certain degree of rotation. The rating of the stepper motor is generally given in steps per revolution of the motor.

The stepper motor used as the drive system for the manipulator must be very accurate so that the manipulator's axes can be positioned accurately. A stepper motor generally gives an error of about 1% per step of the motor. This accuracy is sufficient for accurate axis positioning.

Stepper Motor Types

Three different types of stepper motors are used in robotics applications. These three types are the bipolar, permanent-magnet stepper motor, the permanent-magnet stepper motor, and the bifilar stepper motor. In the following paragraphs, each type of stepper motor is discussed.

Bipolar, Permanent-Magnet Motor.   The **bipolar, permanent-magnet stepper motor** operates on the principle that like magnetic poles repel each other and unlike poles attract each other. Figure 7–13A illustrates the axially oriented permanent magnets located at each end of the axle. The north poles of the permanent magnet are 180° away from the south poles.

The stator poles of the bipolar motor are illustrated in Figure 7–13B. The **stator** is the nonrotating part of the motor that contains the primary (stator) windings of the motor. When a current is passed through these windings, they create a magnetic field in which the permanent-magnet rotor (armature) can rotate.

The stator also has geared teeth, which allow the stepping action to be developed between the stator and the rotor's teeth. The number of teeth on the stator differs from the number of teeth on the rotor. The difference in the number of teeth allows the motor to rotate a predictable amount. That is, the number of teeth on the

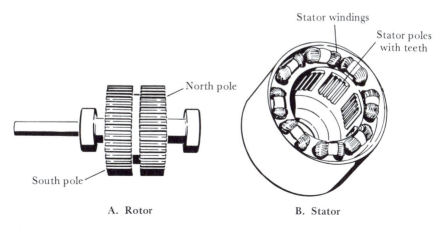

Figure 7–13 Components of a Bipolar, Permanent-Magnet Stepper Motor

rotor and the stator determines the angular rotation of the stepper motor. The greater the number of teeth on both parts, the smaller is the angular rotation of the motor.

In the bipolar stepping motor, a small braking action, or magnetic force, is developed between the permanent magnet and the stator. This small magnetic force produces enough torque to hold the motor in position after the power has been removed. This magnetic field is called **residual torque.**

Permanent-Magnet Motor. The steps in the operation of the permanent-magnet stepper motor are illustrated in Figure 7–14. Like the bipolar stepper motor, the **permanent-magnet stepper motor** operates on the principle that like poles repel and unlike poles attract. In Figure 7–14, the poles mounted on the outside are the stator poles. The stator is made of permanent magnets.

As the stator poles of the motor receive different voltages, their magnetic polarity changes. During the first time period, Figure 7–14A, stator poles A and C are turned on; stator poles B and D are turned off. The applied voltage develops a magnetic field around the stators. The magnetic field causes the rotor to be repelled and, thus, causes the rotor to turn. The rotation of the currents through the stator windings causes the motor to rotate. When the magnetic fields through the stator windings are reversed, the motor rotates in different directions. The number of stator windings on the motor determines the motor's degree of angular rotation.

Bifilar Motor. Different degrees of rotation can be obtained with a **bifilar stepper motor.** The bifilar motor contains the stator wind-

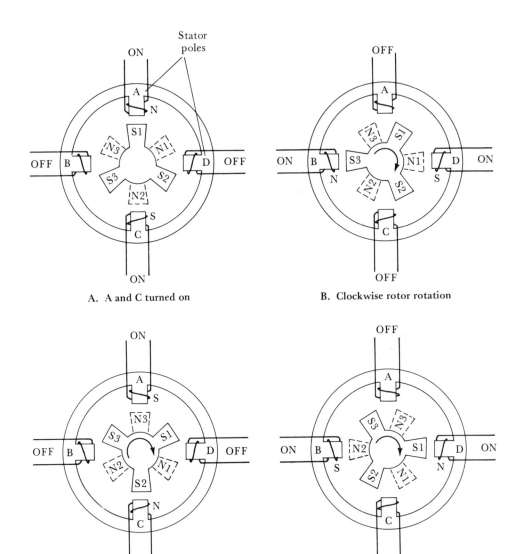

A. A and C turned on

B. Clockwise rotor rotation

C. Change in magnetic fields

D. Continued clockwise rotation

Figure 7–14   Operational Steps of the Permanent-Magnet Stepper Motor

ings. The difference between the bifilar stepper motor and the permanent-magnet stepper motor is that the stator windings are center-tapped in the bifilar motor. In Figure 7–15A, the stator windings L_1 through L_6 are center-tapped. Center-tapped windings provide control of the windings and, thus, produce various angular

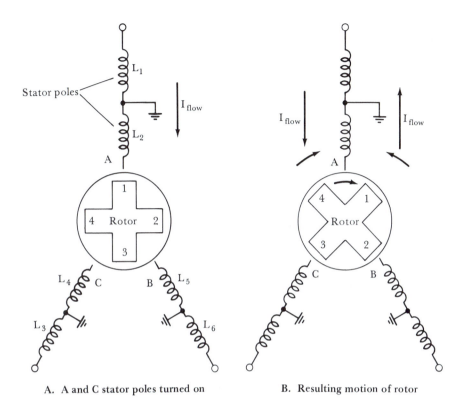

A. A and C stator poles turned on   B. Resulting motion of rotor

Figure 7-15   Operational Steps of the Bifilar Stepper Motor

rotations. When the different stator windings are turned on and off, the rotor is allowed to turn within the stator's magnetic field.

Figure 7-15 illustrates the operation of the bifilar motor. In this motor, two different stator poles come on. In Figure 7-15A, magnetic stator poles A and C are turned on. The magnetic poles attract the rotor so that it turns halfway between the two poles. Figure 7-15B illustrates the resulting movement of the rotor. This method of operation gives greater positional accuracy to the bifilar motor and greater accuracy to the placement of the rotor between the stator poles of the motor.

≡ DC BRUSHLESS MOTORS

All of the motors discussed so far are motors that require brushes for the switching of the armature currents. Because of the mechanical switching, the brushes tend to wear and then cause arcing in the motor. Thus, with these motors, the robot may become unsafe in work areas that contain explosive fumes. Also, the wear generated by the brushes causes brush dust to develop. This dust gets into the bearings of the motor and causes damage to the system.

Instead of brushes, solid-state electronic devices can be used to switch the armature current. Electronic switching is called **commutator action**. A motor that uses electronic switching is called a **brushless DC motor**. Two types of brushless motor are discussed here: split-phase, permanent-magnet motors and Hall effect motors.

Split-Phase, Permanent-Magnet Motors

Figure 7–16 shows the transistor-switching circuitry used in brushless DC motors. The motor in the figure is a **split-phase, permanent-magnet motor** with center-tapped windings. The switching action for the motor is developed by the oscillator circuit consisting of transistors Q_1 and Q_2. The frequency at which commutator action is developed is determined by the resistance, the capacitance, and the inductance of the windings of the motor.

The main disadvantage of the brushless split-phase motor is the low torque developed for driving loads. Also, this motor develops very large starting torques. However, gearing can be employed with the motor to improve the motor's ratings. The main advantage of brushless split-phase motors is that they have no brushes. Thus,

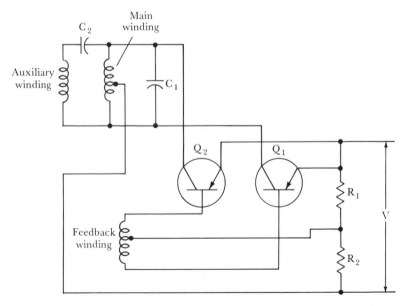

Figure 7–16 Transistor Switching in a Split-Phase, Permanent-Magnet, Brushless DC Motor

the motor has a potentially longer life than the brush motors, and it presents no explosion hazards.

Hall Effect Motors

The **Hall effect motor** has been developed over the years as a long-lasting, high-efficiency, high-reliability, and low–power consumption motor. Hall effect motors are generally found in low-technology robots. The principle of operation of the Hall effect motor is current passing through semiconductor material to generate a voltage. The direction of the current flow through the semiconductor determines the magnetic field around the semiconductor.

Figure 7–17 illustrates the Hall voltage V_H and the magnetic field B developed. In Figure 7–17A, the control current I_C is passed downward through the semiconductor. A magnetic field B is developed, also in the downward direction. The resulting output voltage V_H has the polarity shown in the figure. In Figure 7–17B, the magnetic field is reversed, as is the output voltage.

Conditions around the semiconductor material are critical in the development of the Hall voltage. Changes in temperature, stress, and current from the DC power supply are all factors that can affect the output voltage.

A typical application of the Hall effect motor is illustrated in Figure 7–18. Two Hall effect generating circuits are employed. There are four stator poles for this motor, P_1 through P_4. On each side of the stator poles are the stator windings, W_1 through W_4. The Hall generators are labeled HG_1 and HG_2. The outputs from the Hall generators are connected to two DC amplifiers A_1 and A_2. Each DC amplifier is responsible for amplifying the output voltage from the generator and sensing the direction of the voltage. When the amplifiers sense the direction of the DC voltage, they turn on the proper set of stator windings.

The permanent-magnet rotor is initially in the position shown in the figure. The magnetic field of the rotor develops a downward force (see Figure 7–17A). This force is sensed by HG_1, which creates a positive Hall voltage. Since no force is detected by HG_2, generator HG_2 remains off. The positive voltage developed by HG_1 is then passed to amplifier A_1, which turns on stator windings W_3. The current flow through W_3 produces a south pole relationship in P_1 and a north pole on P_3. This polarity develops a 90° rotation of the rotor.

As the permanent magnet rotates, the magnetic field now affects HG_2. This action turns on HG_2 and amplifier A_2. The stator windings W_1 are now switched on by the amplifier, and a south pole is established on P_2. Thus, the rotor rotates another 90°. This process

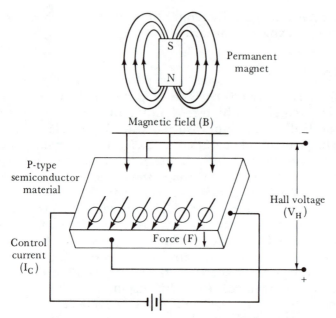

A. Downward direction of current to generator Hall
   voltage and magnetic field

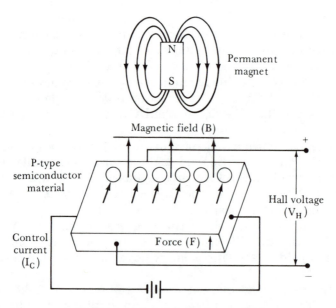

B. Reversed direction of magnetic field

Figure 7–17   Operation of the Hall Effect Motor

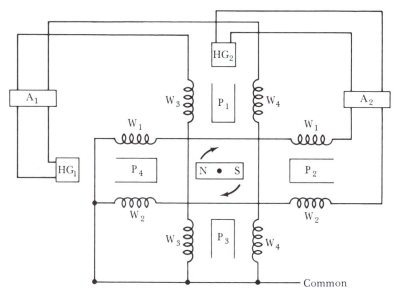

Figure 7–18 Application of a Hall Effect Motor

continues so that the rotor rotates a full 360°. The output of the Hall generating circuit is determined by the position of the permanent-magnet field of the rotor.

The speeds of the Hall effect motors are determined by the value of output voltage from the Hall generator. The greater the output voltage, the faster the motor will rotate. The fluctuation of the Hall output voltage causes the amplifiers in the generator to conduct at different times. Thus, the biasing change on the amplifiers changes the rotational speed of the motor.

If one wants to maintain a constant speed for the Hall effect motor, a constant-feedback circuit can be used. This circuit senses the counter EMF that is developed and makes a comparison. Depending on the results of the comparison, the amplifier can change the amplification factor and thus generate greater output voltages.

AC INDUCTION MOTORS

The **AC motor** has many advantages over the DC motor. For example, the AC motor operates on the alternating current that is supplied by the AC lines in the plant. In contrast, the DC motor requires a conversion of AC to DC power. The AC motors are generally smaller than the DC motors. Yet the AC motors will develop the same horsepower ratings as the DC motors. Also, the AC motor requires no brushes or commutator. Thus, AC motors require less maintenance than the DC motors. Finally, AC motors develop no sparks.

The AC motor most commonly found as a manipulator drive is the **induction motor**. The induction motor is classified into two major types: the squirrel cage induction motor and the wire-wound induction motor. In these induction motors, the stator of the motor has poles for establishing a magnetic field that will rotate. This rotating magnetic field causes the rotor to turn and develop the torque required for moving the load. In the following subsections, we will examine the principles of operation of AC induction motors and the characteristics of the two types of induction motor.

Principles of Operation

The AC induction motor contains two basic components: the stator, which is stationary, and the rotor, which rotates through the magnetic field. The induction motor operates on the basic principle of a magnetic field from the stator inducing a current flow in the rotor. Figure 7–19 illustrates the basic components of the induction motor.

The magnetic field from the stator is induced into a conductor on the rotor. This induced current establishes a magnetic field around the rotor's conductor in the opposite polarity. These opposing magnetic fields repel each other and cause the rotor to develop torque and, thus, motion. The rotor develops motion only in one direction. In order for the rotor to spin, the magnetic field must be moved around the rotor.

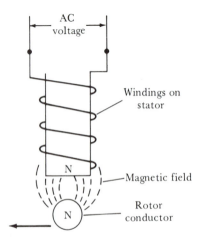

Figure 7–19 Basic Components of an AC Induction Motor

The stator of the induction motor does not move. But if the induction motor is to develop motion, the magnetic field in the stator must rotate. The method used to cause the magnetic field in the stator to rotate is illustrated in Figure 7–20A. In the figure, the stator operates from a two-phase input voltage. This two-phase voltage is connected to the stator poles. Phase V_1 is connected to stator poles N_1 and S_3; phase V_2 is connected to stator poles S_2 and N_4. These two voltages are 90° out of phase. The phase relationship between the voltages is illustrated in Figure 7–20B. When the two voltages are applied to the stator poles, they generate a rotating magnetic field.

Figure 7–21 illustrates the development of the rotating magnetic field in the stator. The two curves for the 90° phase shift of the two voltages are broken down into nine time periods. The rotation of the magnetic field starts at the first time period, T_1. At time period T_1, the voltage from V_1 is maximum, and the amplitude (that is, the value) of V_2 is zero. In this time period, a magnetic field is established between poles 1 and 3. Poles 2 and 4 have no voltage applied and therefore have no magnetic field. At time T_2, voltage V_1 is decreasing in amplitude, and voltage V_2 is increasing in amplitude. Thus, the strength of the magnetic field between poles 1 and 3 is decreasing, and the strength of the magnetic field between

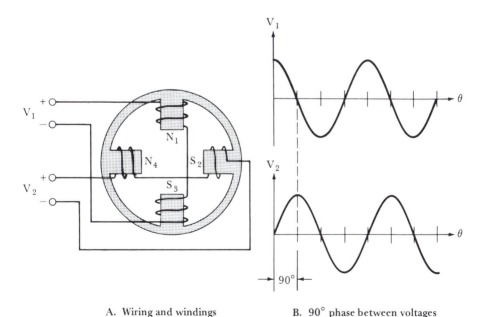

A. Wiring and windings   B. 90° phase between voltages

Figure 7–20   Two-Phase Voltage for Developing a Magnetic Field in a Stator

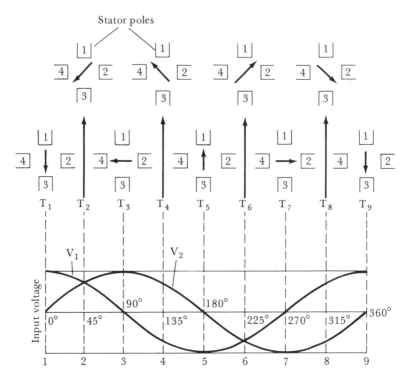

Figure 7–21   Rotation of the Magnetic Field in a Stator

poles 2 and 4 is increasing. The resulting force of the magnetic field
has rotated 45° from the starting point.

At time period T_3, voltage V_1 has reached zero, and voltage V_2
has reached its maximum. This change in voltage develops a change
in the direction of the magnetic field: The magnetic field is now
developed between poles 2 and 4. Since the voltage is zero at the
V_1 input, there is no magnetic field between poles 1 and 3. Thus,
the magnetic field has moved 90° from the starting position.

As the voltages V_1 and V_2 continue to go through their cycles,
the rotation of the magnetic field around the stator poles completes
a 360° movement. Therefore, through the use of a 90° phase shift
in the input voltages, the magnetic field rotates 360° around the
stator.

In order for the rotor to develop motion, it must be constructed
so that it can induce a current from the stator and generate a
magnetic field. Figure 7–22 illustrates a typical rotor used in the
induction motor. This rotor, which is found in the squirrel cage
motor, is constructed of heavy-gage aluminum or copper conductors.
These conductors are connected at both ends by a copper or alu-

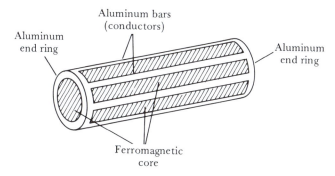

Figure 7–22   Rotor for an AC Squirrel Cage Induction Motor

minum ring. The conductor bars are separated by ferromagnetic material. The ferromagnetic material ensures that the maximum magnetic field developed by the stator will be induced into the rotor's conductors. Remember that this induced magnetic field develops a magnetic field in the rotor, which will then develop the torque necessary for motion of the rotor. Because the conductors are separated by ferromagnetic material, the rotor has been given the name *squirrel cage* rotor.

The AC induction motor has two characteristic features: motor speed and motor slip. The following paragraphs describe these two important features.

Motor Speed.   **Synchronous speed** of the induction motor is defined as the speed value related to the frequency of an AC power line and the number of poles in the stator. Synchronous speed is measured in revolutions per minute (rpm). The synchronous speed is equal to the frequency of the AC input power applied to the stator divided by the number of poles in the stator; this result is then multiplied by 120.

Motor Slip.   In the induction motor, the rotor cannot turn at the same speed as the stator's synchronous speed. So, the rotor and the stator have different speeds. The difference between these two speeds is called the **slip of the motor.** If the synchronous speed of the motor and the rotor speed were identical, then there would be no torque developed in the motor and no relative motion of the rotor. The slip is generated because of the loss of induced current between the stator poles and the rotor's conductors. This small loss of induced current develops a difference in potential between the stator and the rotor and, thus, allows motion of the rotor to develop. Generally, slip is given as a percentage; it may range from 1% to 100%.

When the motor is first started, the slip between the stator and the rotor is the greatest; that is, it is 100%. As the synchronous speed of the motor begins to take over, the slip of the motor is reduced. For example, the slip value might drop to 35%. As the motor reaches its full speed, the slip value might drop to as low as 5%. Remember, though, that in order for the motor to turn, a small amount of slip must be present.

As the rotor turns, the speed of the rotor increases. The buildup of speed also increases the torque of the motor. This buildup of torque increases until the motor's torque matches the torque required to rotate the load. As long as the slip condition exists in the motor, the motor will move the attached load.

Motor Types

As mentioned earlier, AC induction motors are classified into two types: squirrel cage motors and wire-wound motors. The features of each type of motor are described in the following subsections.

Squirrel Cage Motor.   The most common type of induction motor that is used as a manipulator drive is the squirrel cage motor. As we noted earlier, the name for this motor developed from the design of the rotor.

The **squirrel cage induction motor** has different designs to meet the needs of the manipulator drive. The different designs result from changes that are made to the rotor. The rotor changes can affect the torque developed by the motor, the slip of the motor, the speed of the motor, and the maximum current capacity of the motor. The different designs available for the squirrel cage motor are broken down into six classes: class A through class F.

The class A squirrel cage motor is the most popular type; it develops normal torque and starting currents for the motor. The class B squirrel cage motor develops normal torque and low starting currents. The class C motor develops high torque and low starting currents. The class D motor develops very high slip percentages. The class E motor develops low starting torque under normal starting currents. The class F motor develops low torque and low starting currents.

The motor used to drive the manipulator must be able to start, to stop, and to reverse direction constantly. Because of its design, the class D squirrel cage motor is the motor that is best able to handle these demands.

Wire-Wound Motor.   The second type of AC motor found on many manipulator drive systems is the **wire-wound induction motor**. This motor differs from the squirrel cage motor in that the rotor is wound with wire, which is the conductor. Remember that the squirrel cage rotor has bars of aluminum or copper that are used as the conductor.

Figure 7–23 illustrates typical schematic diagrams for the wire-wound induction motor. As shown in Figure 7–23A, the stator windings of the motor have a delta connection. The delta connection of the stator allows three-phase AC voltage to be connected to the induction motor. The rotor connection, shown in Figure 7–23B, is called a wye connection. The ends of the rotor are connected to slip rings, which, in turn, are connected to a variable-resistor bank.

When this motor starts, the rotor resistance is very high, thus producing a very high starting torque for the motor. This large value of rotor resistance is developed from the resistance bank connected through the slip rings. As the motor begins to reach the desired speed, the values for the variable resistors are reduced. At the desired speed, the slip rings remove the resistance bank from the motor circuit. At this time, the motor operates like a squirrel cage motor.

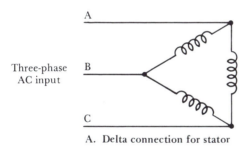

A. Delta connection for stator

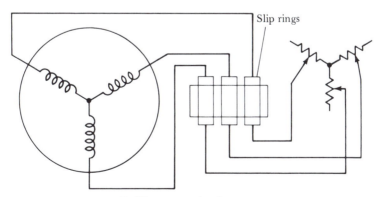

B. Wye connection for rotor

Figure 7–23 Schematic Diagrams for a Wire-Wound AC Induction Motor

The wire-wound motor has several advantages when it is used as the drive system on manipulators. These advantages are smooth acceleration under heavy loads, no overheating of the motor, high starting torque, and good running characteristics. These advantages give the wire-wound motor an edge over the squirrel cage motor in some applications. For example, the ability of this motor to accelerate heavy loads makes this motor useful in gantry mountings or in heavy-lifting operations that must be performed by the manipulator. However, the slow increase in acceleration means that this motor is unacceptable in assembly operations, where quick motion is required.

SUMMARY

In many robots, electric motors are employed as the drive mechanisms. The electric motors used are either DC motors or AC motors.

The electric motor converts electric energy into the mechanical energy the manipulator needs to lift objects. The mechanical energy that the motor develops is called torque.

In many robotic applications, DC motors are used because they operate from 5-volt logic and can be controlled by microprocessors. The major components of DC motors are brushes, a commutator, an armature, and permanent magnets. The brushes supply the main contact of applied DC voltage to the commutator, which, in turn, establishes a magnetic field. This magnetic field opposes the field on the permanent magnets, and the motor then rotates. The types of DC motors used in manipulator applications are series-wound motors, shunt-wound motors, and compound-wound motors. The speed of the DC motor can be controlled in one of three ways: shunt field control, armature control, and solid-state control.

Since the manipulator must be able to reverse its direction, the DC motor must be able to be reversed. Reversing a DC motor can be accomplished by switching the direction of the DC current passing through the motor. The manipulator must also be stopped at certain locations and held in place. The stopping action is controlled by the braking circuits of DC motors.

In many low-technology robots, the stepper DC motor is used to control manipulator movement. The stepper motor operates through DC pulses fed to the motor. These pulses cause the motor to rotate. The rotation of the motor is controlled by the stepper motor's design. In many cases, the stepper motor rotates only one step per pulse.

All of the DC motors described in this chapter develop one

major problem. Over a period of time, the brushes wear and develop electric arcing on the contacts. But with DC brushless motors, arcing is not developed. The main disadvantage of the brushless motors is that they develop very low torque.

The AC motor has several advantages over the DC motor. It delivers high torque, operates without brushes, and operates on the plant's own power.

The AC motors that are used as manipulator drive systems are induction motors. The induction motor contains a stator, which is the stationary section, and a rotor, which is the rotating component. The induction motor induces an AC voltage from the stator into the rotor. This induced voltage generates a magnetic field that is opposite to the stator's magnetic field. This opposition of magnetic fields generates the motion of the rotor and therefore develops the torque of the motor.

The AC induction motors are classified as either squirrel cage motors or wire-wound motors. The squirrel cage motor has bars of conducting aluminum or copper that induce the magnetic field of the stator. The wire-wound induction motor has windings on the rotor that serve as the conductor of the motor.

KEY TERMS

AC motor
armature
armature control
bifilar stepper motor
bipolar, permanent-magnet
    stepper motor
braking
brushes
brushless DC motor
commutator
commutator action
compound-wound motor
DC motor
dynamic braking
electric motor
field coils
Hall effect motor
induction motor
motor reversing

permanent magnets
permanent-magnet stepper
    motor
plugging
residual torque
rotor
runaway condition
series-wound motor
shunt field control
shunt-wound motor
slip of a motor
solid-state armature control
split-phase, permanent-
    magnet motor
squirrel cage induction motor
stator
stepper motor
synchronous speed
wire-wound induction motor

QUESTIONS

1. Define the function of the electric motor as it is used on the manipulator and name the type of power they operate from.

2. State another name for the rotor used in a DC motor.

3. What component of the DC motor connects the DC power to the rotor?

4. Which DC motor develops a speed reduction when the load is increased?

5. What is another name for the rapid acceleration of a series-wound motor?

6. Which DC motor develops constant torque to the load?

7. Describe two methods by which the speed of a DC motor can be controlled.

8. What is the major drawback of armature control for DC motors?

9. How can the direction of a DC motor be changed?

10. Draw a diagram of a circuit using SCRs that will provide directional control of a motor.

11. Which type of braking circuit will reverse the direction of current through the motor?

12. What causes the motion of a stepper motor?

13. What is the major difference between the bifilar and permanent-magnet stepper motors?

14. Discuss the basic principle behind the operation of a Hall generator.

15. Name two disadvantages of the Hall effect motor.

16. What type of rotor is found in a Hall effect motor?

17. What type of conductors are found in a squirrel cage rotor?

18. Which part of the induction motor is stationary?

19. How is the stator connected in a wire-wound induction motor?

20. Define the term *synchronous speed*.

21. The rotor of an induction motor cannot turn at the same speed as the synchronous speed of the stator. What is the difference between these two speeds called?

22. What is the value of the slip when the induction motor is first started?

23. What class of squirrel cage motor develops the greatest amount of slip?

24. Name a general robotics application for the wire-wound induction motor.

25. Name three characteristics of the wire-wound motor.

8

Servo System Control

OBJECTIVES

Upon completing this chapter, you should be familiar with:

— Closed-loop servo system,
— Feedback components,
— Servo amplifiers,
— Programmed servo signals.

INTRODUCTION

The main function of any robotic system is to place the manipulator's end effector at the same location and the same point over and over again. In medium- and high-technology robotic systems, this positioning of the manipulator must be very accurate. Therefore, the drive mechanism must have some type of control system by which positional and velocity axis data can be transmitted to the controller. The controller can then compare these actual axis values with programmed data to ensure that the axis of the manipulator is traveling to the correct programmed position and at the correct velocity.

The motor control circuitry in the robotic system is called the **servo system**. Two basic servo systems are used in motor control: the open-loop servo system and the closed-loop servo system. The closed-loop servo system provides feedback for accurate positioning. Since the accuracy of the manipulator is very important in robotic applications, only the closed-loop servo system is used in robotics. Therefore, in this chapter, only the closed-loop servo system will be described.

In this chapter, the basic operation of the closed-loop servo system will be explored. The operation of the basic components of that system, such as feedback components to identify velocity and position, will be described. The programmed servo signals from the controller are also discussed. Since these signals have a low amplitude, the servo system requires an amplifier to increase the signal strength. Therefore, this chapter also examines servo amplifiers.

☰ CLOSED-LOOP SERVO SYSTEM

A block diagram of a simple **closed-loop servo system** is given in Figure 8–1. The system includes an input comparator, which is responsible for the comparison of the command input signals from the controller's microprocessor and the feedback pulses from the actuator. The servo amplifier is responsible for amplifying the difference between the command pulses and the feedback pulses and feeding these pulses to the actuator. The actuator is the drive system, such as an electric motor, a hydraulic cylinder, or a hydraulic motor. The feedback convertor receives the pulse from the actuator and feeds those pulses back to the input comparator.

In the following subsections, we will examine the closed-loop servo system in detail. In particular, we will consider its principles of operation, the definitions of some basic terms, and the system's major components.

Principles of Operation

An example of a simple servo operation is illustrated in Figure 8–2. In this example, the speed and the position of the motor are controlled by a feedback network. The feedback network is a visual observation by a human. The speed of the motor is controlled through the variable resistor. As the resistance of the variable resistor is changed, the motor's speed will decrease or increase. For example, as the motor speed increases, the operator can increase the resistance of the variable resistor, thus causing the motor to slow down. A hold button allows the operator to stop the motor when the end effector has reached the desired position.

In this simple example of a closed-loop servo system, the rotation and the velocity of the servo motor are controlled through visual feedback from the operator. The feedback in this case goes to and from the human operator of the motor. So that a system may

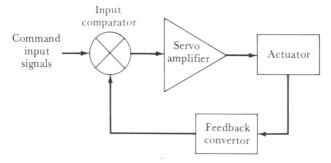

Figure 8–1 Block Diagram of a Closed-Loop Servo System

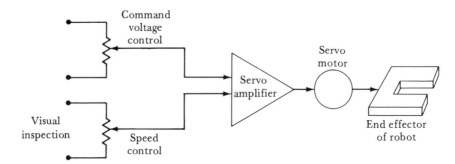

Figure 8–2 Simple Servo Operation

run without a human operator, electronic circuitry is built into the servo system to control the feedback, the programmed pulses, and other components that are necessary for complete operation.

In a closed-loop servo operation, all the positional data and the velocity of axis travel can be monitored. These axis parameters are sent back to the servo's monitoring device, where the data is compared with programmed data. The feedback signal is sent through **transducers**, which are devices that convert one form of energy into another form of energy. The monitoring function is completed by the microprocessor located in the robot's controller.

Definition of Terms

Many terms connected with the operation of servo systems should be understood before we look at the components of a closed-loop servo system. The following terms are common only to closed-loop operation:

— **Frequency response**: A measure of how well the servo system is able to respond to sharp changes in direction or in velocity.
— **Feedback**: A process used to send the positional data and the velocity of the axis to the robot controller.
— **Following error**: The difference between the actual speed of the actuator and the commanded speed of the system.
— **Gain**: The amount of change developed between the input and the output signals in the servo system.
— **Instability**: The axis oscillation caused by the servo hunting for its position. Instability generally arises from a poorly designed servo system or a misaligned servo system.
— **Sensitivity**: The amount of signal that must be generated before the servo system can overcome friction and move the axis.

— **Velocity error**: A difference between the programmed velocity and the present velocity; this difference develops axis movement.

Components

Figure 8–3 illustrates an expanded servo loop for measuring the velocity and the position of an axis. In this block diagram, several different components are included in the overall servo system in order to generate the feedback data required.

The input of the servo system shown in Figure 8–3 is from the robot controller's axis control circuitry. Positional data that has already been programmed is sent to the input position comparator circuitry. The amplifier that follows the position comparator increases the strength of the compared signal and sends it to the speed comparison circuitry. In the speed comparison circuitry, the feedback signal of the velocity of the axis is compared with the programmed speed. If there is any difference between the programmed speed and the actual speed, the servo will correct the velocity. The speed comparison circuitry then feeds an amplifier circuit. Again, the amplifier only conditions the signal to control the servo mechanisms.

The servo control device in Figure 8–3 can be a hydraulic servo valve or an electronic control, such as an SCR. The pulses from the amplifier are sent to the servo control circuitry. The servo control circuitry directs the motor (hydraulic or electric drive) to rotate. The motor then moves the axis.

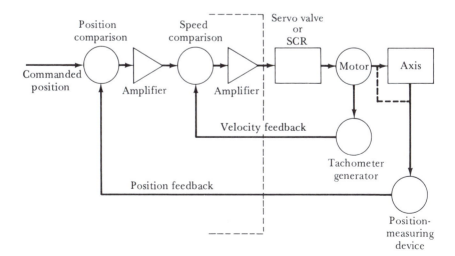

Figure 8–3   Servo Loop for Measuring the Velocity and the Position of an Axis

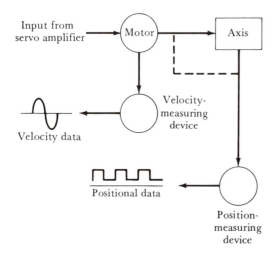

Figure 8–4 Waveforms for Velocity and Position

    Two feedback components are used in the servo operation of Figure 8–3. Both components are transducers that convert one form of energy to another form. The tachometer generator converts the velocity of the actuator into electric pulses. This operation gives the velocity feedback. The position-measuring device is an encoder that converts the angular position into positional feedback signals. Both of these transducers convert feedback signals into electric pulses. These pulses are then fed back to the robot controller's axis control circuitry, where they are compared with the programmed data. Any errors, between these two signals, that are detected are then corrected by the servo system.

    The feedback signals are shown in Figure 8–4. The two signals are the positional feedback signal and the axis velocity. The positional data is collected from a positional transducer that converts angular or linear positional data into electronic pulses. The positional transducer may be an optical shaft encoder or a resolver. The velocity transducer used for monitoring the speed of the axis is a tachometer generator. These two components work together to measure the axis speed and position. These components will be discussed in detail in the next section.

FEEDBACK COMPONENTS

The **feedback transducer** converts hydraulic pressure or the mechanical rotation of the system into an electric signal. This electric signal is then sent to the robot controller, which makes adjustments so that the robot manipulator can be accurately positioned.

The feedback transducers can be described by the following characteristics:

— **Resolution**: A measure of the ability of the transducer to break down the angular position into fineness of detail in a spatial pattern.
— **Accuracy**: A measure of how close the actuator can come to the commanded position.
— **Repeatability**: A measure of the ability of the transducer to give the same reading when returned to the same position.

The feedback transducers used in the robotic system are either digital or analog devices. The digital devices convert the velocity of the axes into digital signals. The analog devices convert the angular rotation of the axes into an analog signal. The signals, digital or analog, are then fed back to the controller. In the majority of medium- and high-technology systems, digital output transducers are used. The main reason digital devices are used is that the internal operations of the controller are digital signals (either in an on or off state).

Two types of transducers, velocity and positional, are used to convert the feedback signals into usable signals in the servo comparator circuitry. The velocity transducer converts the speed of the axis into an electronic signal, and the positional transducer converts the angular or linear position of the axis into an electric pulse. Both types of transducer are discussed in the following subsections.

Velocity Transducers

The **velocity transducer** is generally a tachometer generator. The **tachometer generator** takes the speed of the axis and converts it to a proportional voltage level. The most common types of tachometer employed produce an AC or a DC signal at the output.

Figure 8–5 illustrates the typical output of a tachometer generator employed as a velocity feedback component. The graph relates the speed of the device to its output voltage. As the revolutions per minute of the tachometer generator increase, the output voltage also increases. For an example, this tachometer generator is rotating at 1000 rpm, which equals an output voltage of 20 volts DC.

Another velocity transducer is the **linear induction transformer**. This component consists of a magnet and a coil. A voltage is generated at the output of the coil. This voltage is proportional

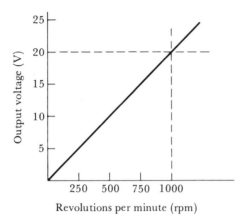

Figure 8–5 Output Voltage versus Speed for a Tachometer Generator

to the speed at which the magnet moves through the coil. The linear induction transformer is primarily used in hydraulic servo systems.

Positional Transducers

Positional transducers convert the angular or linear motion of the axis into an electric signal. Again, this signal can be either a DC or an AC signal, and it can be either an analog or a digital signal. The type of signal outputted depends on the operation of the controller.

Positional transducers are of three types: variable transformers, resolvers, and optical shaft encoders. Each of these types is discussed in the following paragraphs.

Variable Transformer.   A block diagram of the **variable transformer** is presented in Figure 8–6. In this application, a hydraulic servo system is being positioned. The primary of the variable transformer is connected to an AC input. Two secondaries of the transformer are connected in a series-opposing manner. The core of the transformer is a high-permeability core attached to a plunger, which acts as the coupling between the primary and the secondary of the transformer. When the plunger is in the center of the core, the voltage induced from the primary to the secondaries is equally distributed between the two secondaries. Since the two coils are series opposing, their voltages are opposite and equal and will cancel. Therefore, the net voltage supplied to the demodulator is zero. So, the output to the servo valve is zero, and the servo valve maintains the pressure of hydraulic fluid through the control valve. The position of the system is stable.

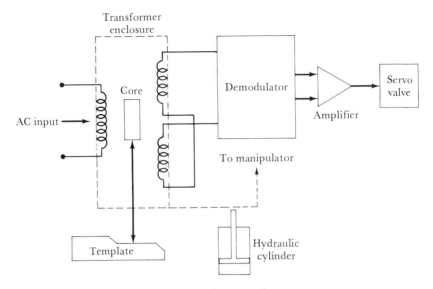

Figure 8–6 Block Diagram of a Variable Transformer

As the manipulator axis moves, the template connected to the core will move. This template movement causes the transformer core to move. The transformer is then in an unbalanced condition. That is, the voltages of the two secondaries of the transformer are unequal. This unbalanced condition is detected by the demodulator, which produces a DC voltage to the servo amplifiers. The amplified error signal is then sent to the servo valve, which will activate an increase in pressure to the hydraulic cylinder. This increase in pressure causes the cylinder to move. Again, the core of the transformer is moved, causing a balance condition in the transformer. The balanced condition means the template is in position.

Resolver.   The **resolver** is a positional transducer that converts the angular position of the axis that is inputted to it into a proportional output voltage. Figure 8–7 illustrates the windings for a typical resolver. The input voltage is applied to the rotor windings, and two stator coils provide the output. These output coils are 90° apart, and hence they will generate various voltage levels, depending on the position of the resolver's rotor. The outputs of the resolver are then fed into a high-gain amplifier, where the two signals are compared. Their difference is outputted to the robot controller's servo system.

Optical Shaft Encoder.   Figure 8–8 illustrates the **optical shaft encoder**. The optical encoder is made up of four basic components, as

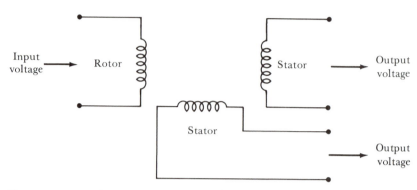

Figure 8–7   Resolver Windings

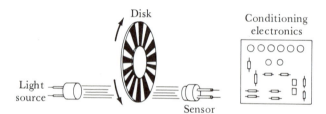

Figure 8–8   Components of an Optical Shaft Encoder

shown in the figure. The light source can be an incandescent lamp or a light-emitting diode (LED). The disk has alternating opaque and translucent patterns. The sensor senses the light (or no light) passing through the disk. Finally, the conditioning electronics is circuitry that senses the difference between the light and the dark areas and converts these signals into usable data. The resolution of this type of transducer can be increased by adding another disk. The second disk allows only the light areas to be transmitted to the sensing element.

Optical encoders can be either incremental encoders or absolute encoders. Each type is described in detail in the following paragraphs, because the optical encoder is the most common type of transducer used in robotics.

The **incremental encoder** can give the controller both positional measurements and velocity measurements. The incremental encoder is connected to the motor of the axis. The rotation of the motor causes the disk to rotate. The rotation of the disk generates a series of digital pulses as output. These pulses are then fed to an electronic counter that keeps track of the number of pulses generated from the disk. The incremental encoder has a calibration mark that tells the digital counter when to store the pulses. The counter then supplies this information to a comparator circuit that compares the actual position with the programmed position. If there is an error generated, the servo corrects the difference.

The positional data stored in the counter can be lost if the power fails in the robot controller. If the power fails, the system will have to be reset. The resetting process is generally accomplished through the zero-return operation, which returns the manipulator to a preprogrammed position.

The accuracy and resolution of the incremental optical encoder system can be increased by the addition of a **multiplier circuit**. Figure 8–9 illustrates such an incremental encoder system. In this example, the axis of the robot is required to move a certain distance. The pulse output is developed by the rotation of the shaft of the encoder. Each time the shaft rotates one cycle, the encoder generates 2500 pulses. These pulses are then passed through a multiplier circuit. The multiplier circuit in Figure 8–9 multiplies the pulse count by 4. The number of pulses being fed back to the digital counter is now 10,000 pulses instead of 2500. Therefore, the accuracy of the counter is 0.0001 per inch of travel. The number of cycles generated for one revolution of the encoder can be up to 5000 pulses, which, when passed through a multiplier, will increase the accuracy of the encoder system.

The **absolute encoder** generates a parallel output of digital code to the counter circuitry. Figure 8–10 illustrates an absolute encoder system. The ten different channels around the disk are shown. Each channel is comprised of a translucent and an opaque section. As the disk rotates around the shaft of the motor, the light passing through the disk produces a digital code. The digital codes comprise a parallel output to the counter circuitry. The parallel output code gives the controller the positional information.

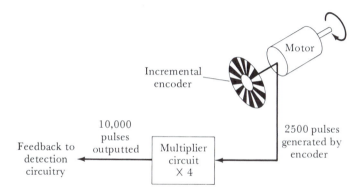

Figure 8–9   Feedback Pulses from the Encoder to the Multiplier Circuitry

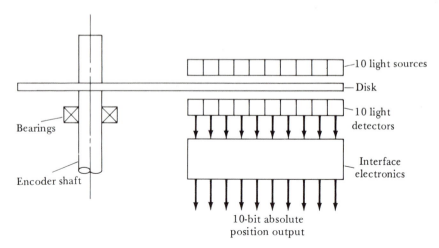

Figure 8–10   Absolute Encoder for Producing a 10-Bit Output Signal

The code generated as output is either a binary code, a BCD code, or a Gray code value. The following table lists the Gray code generated from an absolute encoder:

Decimal	Gray code
0	0000
1	0001
2	0011
3	0010
4	0110
5	0111
6	0101
7	0100
8	1100
9	1101
10	1111
11	1110

The Gray code changes from one number to the next by changing only one bit in the code. For example, notice the difference between the decimal numbers 6 and 7. The only bit that changes is the last bit. Changing only one bit at a time ensures the accurate positional information required by the servo system. Also, the absolute encoder system does not have to be calibrated every time power fails.

In many high-technology manipulators, RAM memory is connected to the encoders. The pulses generated by the encoder are stored in RAM memory. Therefore, the encoder will remember pulse count, which is important for positional data. Because the RAM memory will lose its data when power fails, batteries are connected to the memory power source. Therefore, when power fails, the batteries turn on and supply the memory with power. The memory then can return the position of the axis. When total power is restored, the batteries are automatically switched off, and the stored memory is processed. This allows for the manipulator's axis to retain its positional data.

SERVO AMPLIFIERS

The transducers used for positional and velocity feedback require some type of drive circuitry to control the speed and the direction of the motor. This driving operation is handled by the **servo amplifier**, which takes the signal supplied by the controller and drives the servo motor.

The servo amplifier system controls either a DC servo system or an AC servo system. These two amplifier systems are described in this section.

AC Servo Amplifier

A **solid-state AC servo amplifier** amplifies the positional error voltage to produce the motor control voltage. Figure 8–11 gives a block diagram of an AC servo amplifier. A DC voltage, V_{in}, is fed to an

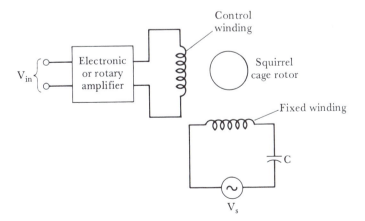

Figure 8–11   Solid-State AC Amplifier

electronic amplifier, which supplies the stator windings of the motor. An AC signal, V_s, is supplied to the fixed windings of the motor. The alternating current supplied to the fixed windings passes through a capacitor, which causes the current to be shifted 90°. The objective of this amplifier circuit is to synchronize the voltages V_{in} and V_s.

As the error between the actual position of an axis and the programmed position increases, the error voltage also increases. This increase, in turn, causes V_{in} to increase, which produces a larger current at the output of the amplifier. This increase in current supplies the stator windings with more current, thus causing the motor to rotate faster. As the motor's speed increases, the error voltage is reduced. Eventually, the motor will maintain the programmed speed.

As the motor reaches the programmed position, the error between the programmed axis position and the physical location of the axis decreases. This decrease causes the following error to decrease, and the motor will slow down as the axis approaches its programmed position. Once the axis has reached its programmed position, the error is zero, and the motor will stop.

The basic requirements of the AC servo amplifier are the same as the requirements of any amplifier circuit used in industrial applications. That is, the amplifier should have high input impedance so that the input source is not loaded down. The amplifier should have large voltage gains that are independent of the temperature. The amplifier should have low output impedance so that the amplifier can supply the servo motor with the required current. Finally, the amplifier should operate within the temperature range so that it does not overheat.

Figure 8–12 illustrates a typical four-stage transistor amplifier used as an AC servo amplifier. The error signal is generated in a frequency-to-voltage converter. The error signal is supplied to the base of Q_1, the input transistor. The error signal is then passed through Q_2, which is a voltage amplifier. The signal continues on to be processed through Q_3, which is connected to the primary circuit of transformer T_1. This circuit is a tuned circuit and passes only 60 hertz to the secondary. This 60-hertz signal biases transistors Q_4 and Q_5 and causes conduction through the primary of transformer T_2. The current flow through the primary produces a secondary voltage that is supplied to the control windings of the servo motor. This input to the servo motor causes it to rotate. The rotating motor then develops the axis motion.

As Figure 8–12 shows, the servo motor's control windings receive the error signal, but the servo motor's fixed windings are supplied by an alternating current that has been shifted by 90°.

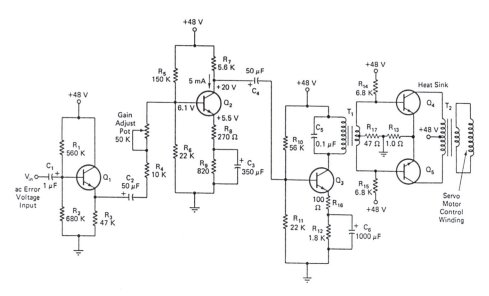

Figure 8–12   Four-Stage Transistor AC Amplifier

The error signal fed to the control windings will regulate the motor's rotation, clockwise or counterclockwise. If the error signal leads the fixed-windings current by 90°, the motor will rotate in one direction. If the error signal lags the fixed-windings current by 90°, the motor will rotate in the opposite direction.

DC Servo Amplifier

The **DC servo amplifier** is generally found in applications where high current and heavy loads are required. Figure 8–13 illustrates one basic type of DC servo amplifier. This amplifier consists of SCR circuitry. The two SCRs are used to directionalize the motor. In the circuit of Figure 8–13, SCR_1 is used to control the motor in a clockwise direction, while SCR_2 is used to control the motor in a counterclockwise direction. The supplied AC voltage V_{AC} gives the motor the stator current necessary for rotation of the motor. The direction of current flow depends on which of the SCRs is conducting.

The error signal V_e is applied to the preamplifier circuitry. The amplified error voltage is developed at the output of the preamplifier. This signal is V_{out}. The signal is then rectified by diode D_1 and applied to the anode of the four-layer diode D_2.

The four-layer diode D_2 will conduct at different levels of error signal. If the error signal is large, the four-layer diode conducts

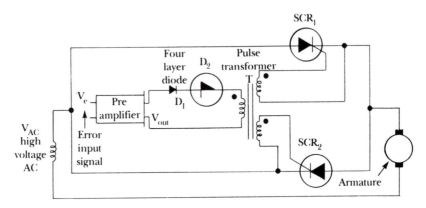

Figure 8–13   DC Amplifier Using SCRs

early and supplies a burst of current to the primary of the pulse transformer T. The pulse transformer then supplies the secondaries of T, which, in turn, supply the gates of each SCR.

The pulsed voltage from T turns the SCRs on. But only one of the SCRs will conduct, because the SCRs are properly biased at that moment. Let us say that SCR_1 turns on. When SCR_1 is on, current flows through the armature of the DC servo motor. This current flow causes the motor to rotate in one direction. If, on the other hand, SCR_2 is turned on, then the servo motor will rotate in the opposite direction.

If the error signal is small, the same series of events occurs, except that the four-layer diode will not turn on right away. Thus, the firing of the SCRs is delayed.

Figure 8–14A illustrates another type of DC amplifier, a transistor bridge servo control. In this circuit, four transistors, Q_1, Q_2, Q_3, and Q_4, are used to control the speed and the direction of the DC motor. The transistors are biased by the DC power supply, V_{CC}. This DC voltage is supplied to the armature of the DC motor. This voltage is guided by the transistor conduction path and establishes the direction of the motor's rotation.

The transistors are connected as a bridge circuit, and the motor is connected between the emitters of transistors Q_2 and Q_4. The bases of the transistors are connected to four pulsed, digital control signals. When a signal is applied to the transistor base, the transistor will conduct.

Figure 8–14B shows what happens when pulsed signals are applied to Q_1 and Q_4, while Q_2 and Q_3 are turned off. The signals applied to Q_1 and Q_4 cause current to flow from ground through the

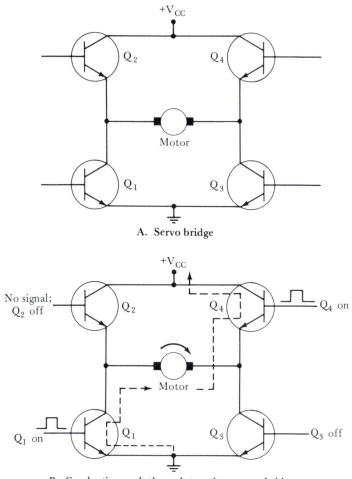

A. Servo bridge

B. Conduction path through transistor servo bridge

Figure 8–14 Transistor Bridge DC Amplifier

transistors. The current flowing through the motor's windings causes the motor to rotate in one direction. An increase in the signal supplied to the base of the transistor will cause the transistor to conduct more current. When conduction increases, the motor rotates faster. As long as the signals are applied to Q_1 and Q_4, the servo motor will rotate in one direction.

When the pulses are applied to Q_2 and Q_3, the motor will rotate in the other direction. Thus, the pulses applied to the base of the transistor bridge circuit control the direction of the servo motor. As in the other servo amplifier circuits, the feedback error signal is detected, amplified, and distributed to the various transistors in the bridge circuitry.

PROGRAMMED SERVO SIGNALS

Many high-technology controllers have microprocessor-based servo systems. These microprocessor systems use **programmed servo signals** from the positional data of the controller to move an axis to the desired position. Constant feedback from the servo is sent to the servo's microprocessor for constant updating of the position of the axis. The travel of the axis can be monitored by the servo system, but in many high-technology controllers, an alarm circuit is also built into the controller's microprocessor system. This alarm system detects error conditions, such as counting pulses between the programmed position or the axis traveling at a speed greater than the programmed speed.

Positional signals and servo alarm conditions are described in the following subsections. Note that the microprocessor-based servo system operates just like the microprocessor systems discussed in Chapter 3.

Position and Movement Signals

The programmed servo signals are generated from the microprocessor in the controller. These programmed signals can also be generated from the teach pendant when the operator requires movement from the axes of the robot, or they can be generated from the positional data stored in the controller's memory. The programmed pulses are stored and outputted from the memory as digital signals. They must be directed to the correct location for driving the servo motors.

Figure 8–15 is a block diagram of the operation of moving an axis from point A to point B to point C. Each of these positions requires an address block that will command the axis to move. The main microprocessor in the controller is responsible for developing the control of axis movement and also for directing the axis command signal to the correct axis. Each axis of the manipulator requires some type of servo system for axis motion.

The servo's microprocessor memory circuit has two separate storage locations, called *registers,* for positional data. One register is the command register, and the other is the actual-position register. At the start of a new positioning operation, the coordinates for the new point are read into the command register from the master program memory. The actual position of the axis is stored

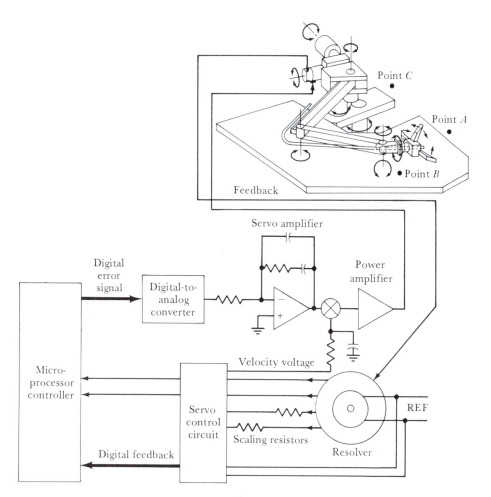

Figure 8–15   Microprocessor Control of a Robot's Axis

in the actual-position register. A third register stores the command velocity for positioning the axis.

During a movement of an axis, the microprocessor samples the two registers that contain positional information. That is, the microprocessor scans and compares the positional information in the actual and the command registers. The difference between the data in these two registers is then sent to the velocity register to develop the velocity command signal for axis movement.

The velocity command signal is sent to the velocity command printed circuit board. This board supplies command voltages to the servo motor. For instance, the command voltages might be supplied to the bases of the transistors in the transistor bridge circuitry.

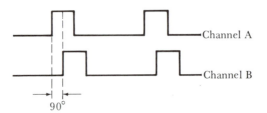

Figure 8–16   90° Phase Shift in the Resolver's Output Signal

As the microprocessor continues to compare the two registers containing command position and actual position, error signals are generated and sent to the velocity register. An increasingly smaller error between the data in the command and the actual registers indicates that the servo motor is reaching its programmed position. At the programmed position, the axis drive motor stops and signals the microprocessor that the robot axis is in position. The robot manipulator then calls up the next positional data and repeats this process.

The positional information of the axis is developed from the resolver. The resolver generates two square waves at the output, one from channel A and one from channel B, as shown in Figure 8–16. These two waveforms are 90° out of phase. The resolver sends these waveforms back to the microprocessor circuitry for positional comparison. The command signal also generates square waves, and these signals—the resolver waveforms and the command waveforms—are then compared. The difference between these signals develops the command signal for axis drive.

Alarm and Monitoring Signals

In many high-technology controllers, protection devices, called **alarm conditions**, are built into the program for the safety of the robotic system and the worker. These alarm conditions relate to the servo systems. The alarms are generated when a manipulator axis continues to move after it has reached the programmed position. An error, called the following error, is generated because the data in the actual register and the command register are equal, indicating that the axis requires no additional movement. But the servo system is continuing to send back positional updates, and these updates are changing because the axis is moving. This condition develops when the servo motor tries to move more than the command signal specifies. A preprogrammed **interrupt routine** in the microprocessor is started once the alarm condition signal has been given. The interrupt routine causes the microprocessor to shut down and stop the robot's operation.

Additional monitoring circuits in the servo section of the axes give alarm signals for velocity errors, motor overheating, high voltage to the motor, or high current to the motor. In each of these alarm conditions, the servo system will shut down, and the axis of the robot will not move.

SUMMARY

Servo systems are used in many manipulator drive systems. The servo systems are of two types: the open-loop servo system and the closed-loop servo system. The closed-loop servo system, which provides accurate positioning of the manipulator, detects the velocity of the axis as well as the angular or linear position of the axis.

Several common terms describe the operation of the closed-loop servo system. The frequency response of the servo system is a measure of how well the servo will respond to changes in direction of the actuator. The feedback signal gives the positional data or the velocity of the servo system. The gain of the system is its amplification. The following error of the servo system is the difference between the programmed position and the physical location of the axis. The instability of the servo system describes how much a system will search for its position. The sensitivity of the servo system is the amount of signal required to cause the servo system to operate. The velocity error of the servo system is the difference between the programmed velocity and the actual velocity of the servo system.

The components of the servo system detect and correct the velocity difference between the programmed velocity and the actual velocity of the axis. Detecting positional information is also the function of the servo system. Positional differences between the programmed position and the actual position of the axis are corrected by the servo system.

Feedback signals from the velocity and positional transducers are converted into electric pulses and fed back to the servo system's microprocessor for comparison. Feedback transducers must develop the correct information for the servo system. So, the feedback transducer must have resolution, accuracy, and repeatability. The feedback transducer converts analog signals into digital signals that are fed back to the controller.

The velocity transducer is a tachometer generator that converts the velocity of the axis into an analog signal that is detected by the servo system and identified as the speed of the axis. Positional transducers determine the position of the axis. These transducers can be rotational transducers or linear transducers. Positional transducers can be resolvers, variable transformers, or optical shaft encoders.

The optical shaft encoder is the device commonly used in robotic servo systems. It is a disk made up of alternating opaque and translucent patterns. Coupled with a light source, the alternating patterns will produce a series of digital pulses. These pulses are then counted by the servo system so that the servo can identify the actual position of the axis and compare it with the programmed location.

Two types of optical encoder are used in robotic systems: the incremental encoder and the absolute encoder. The absolute encoder produces a parallel output of digital pulses. The incremental encoder produces a series of digital pulses. The absolute encoder system produces an output signal in a BCD code, a Gray code, or a binary code.

The servo amplifier in servo systems is responsible for increasing the level of the signal supplied to the servo motor. The amplifier can amplify either a DC or an AC signal. Many servo amplifiers are solid-state amplifiers that provide high gain factors as well as high current output for driving the actuator.

KEY TERMS

absolute encoder
accuracy
alarm conditions
closed-loop servo system
DC servo amplifier
feedback
feedback transducer
following error
frequency response
gain
incremental encoder
instability
interrupt routine
linear induction transformer
multiplier circuit

optical shaft encoder
positional transducer
programmed servo signals
repeatability
resolution
resolver
sensitivity
servo amplifier
servo system
solid-state AC servo amplifier
tachometer generator
transducers
variable transformer
velocity error
velocity transducer

QUESTIONS

1. What is the major difference between the open-loop and the closed-loop servo systems?

2. Name the two parameters of the manipulator's axis that the servo system is able to control.

3. Name the characteristic of the servo system that describes how well the servo can respond to sharp changes in direction.

4. What condition can arise from a poorly designed or a misaligned servo system?

5. Servo system A requires 10 volts of input to activate the system; servo system B requires only 10 millivolts. Which servo system has the better sensitivity, A or B? Explain why.

6. Which device in the servo system receives an input signal from the controller and the position-measuring device?

7. What circuit is the output of the speed comparison circuitry fed to?

8. Which type of transducer converts the angular position of the axis into electric pulses?

9. What characteristic of a feedback transducer describes its ability to convert a sample signal into a readable signal by the servo system?

10. What types of signals does the feedback transducer generate?

11. What type of velocity transducer converts the speed of the axis into an electric signal?

12. What type of voltage is generated from the output of the velocity transducer?

13. For what type of robotic system is the linear induction transformer generally used as a positional transducer?

14. What is the phase difference between the stator and the rotor coil on a resolver?

15. Name the four components that make up an optical shaft encoder.

16. What two types of data does the incremental encoder supply to the servo system?

17. How many pulses are generated in each revolution of the incremental encoder?

18. Describe how the accuracy of an encoder system can be increased.

19. What type of encoder system gives a Gray code output?

20. Why is the Gray code encoder used as a positional encoder in many feedback systems?

21. What signal does the servo amplifier increase in strength?

22. Name three requirements of a servo amplifier system used in industrial robotic systems.

23. For the circuit of Figure 8–13, what determines the time at which the four-layer diode conducts?

24. During the movement of an axis, what two registers does the microprocessor sample to identify the position of the axis?

9

Robotic Gears and Linkages

OBJECTIVES

Upon completing this chapter, you should be familiar with:

— Basic concepts of mechanics,
— Gears,
— Belts,
— Chains.

INTRODUCTION

The drive system of the robotic manipulator is composed of motors (or actuators), gears, and linkages. The gears and linkages transfer the energy developed by the motor to the manipulator so that it can perform work.

The robotic manipulator is nothing more than a series of mechanical arms. The arms, through the use of drive systems (electric, hydraulic, or pneumatic), are connected together to produce work. The axes' mechanical operation is driven by the motor, or the actuator. That is, the actuator transfers its energy to the manipulator's axis.

The transfer of energy is accomplished through three basic means: the gear (the most popular method), the belt, and the chain. The gearing can be arranged in several different patterns in order to get the correct ratio of motor speed to axis movement speed. For instance, motor speed can be reduced to the controlled speed of the axis through a component called a harmonic drive.

In this chapter, the gearing and linkages that are used for the movement of the manipulator will be explored. The basic concepts of mechanics will first be discussed since these concepts are needed in order to understand the transfer of energy from the actuator to the manipulator.

BASIC CONCEPTS OF MECHANICS

Mechanics is the study of the operation of different forces that act on a body. This science is broken down into two basic areas: statics and dynamics. **Statics** deals with forces and their effect on a body at rest. **Dynamics** is the study of force as it works on a rigid body in motion. In the field of robotics, dynamics is further broken down into two additional classifications: kinematics and kinetics. **Kinematics** is the study of the motion of a rigid body without consideration of the forces acting on that body. **Kinetics** is the study of the forces on a rigid body in motion and the effects the forces have on that body.

All of these areas of mechanics are involved in the operation of a robotic system. That is, the robot manipulator is a rigid body that must be placed in motion. The force that is required to move the manipulator must be transferred from the actuator to the manipulator. This transfer is accomplished through the use of chains, gears, or belts. The amount of torque that is developed by the actuator can be increased or decreased through the ratio that is established between the varous linkages on the manipulator. So, the transmission of energy, the work of the manipulator, the power of the manipulator, and energy and torque are all important mechanical concepts for robotics. Hence, these topics will be discussed in the following subsections.

Transmission of Energy

In most robotic systems, the point where the motor (or actuator) is connected is generally not the point where work is done. For example, in Figure 9–1, the drive motor for the robot is connected at the left end of the arm. Through the use of gears, the motor's energy

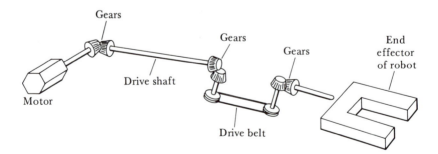

Figure 9–1   Transfer of Energy from the Motor to the End Effector

is transferred to the drive shaft and then to the other end of the arm, where the work is to be done by the end effector.

In many robots used in the industry today, the transfer of energy is a mechanical transfer. **Mechanical transfer of energy** is controlled through the use of levers, chains, pulleys, belts, cams, and gears. The energy is transferred from the actuator to the point at which the work will be done.

Some robotic systems use other methods for transferring energy, of course. For example, *electrical transfer of energy* is accomplished through electric wires to the point where the work will be done. *Pneumatic transfer of energy* is accomplished through the use of compressed air, which is transferred through piping to a pneumatic actuator that will accomplish the work. *Hydraulic transfer of energy* is accomplished through the use of pressurized fluids piped to an actuator. The actuator will then cause work to be done.

The work done in a robotic system may, for example, be the lifting of a part from one location to another or the moving of an arc-welding gun down a welding path. The key in the transfer of energy is the actuator (or motor). The actuator converts the electric, pneumatic, or hydraulic energy into the mechanical energy required to perform the work.

Work of the Manipulator

The **work** performed by the manipulator is defined as the movement of an object through a distance. The amount of work done is measured in foot-pounds (ft-lb).

The work done can be determined through an equation. This equation uses the factors of the weight (force exerted) that is moved and the distance the weight is moved. The equation for work is as follows:

$$W = D \times F \tag{9-1}$$

where

W = work done, in foot-pounds
D = distance the object is moved, in feet
F = force exerted, in pounds

The use of Equation 9–1 is illustrated in the following example.

Example
A robot manipulator is to raise a 110-pound die from a press. The distance the die casting must move is 15 feet. Find the amount of work the manipulator provides.

Solution
Use Equation 9–1, substitute, and solve:

$$W = D \times F = 15 \text{ ft} \times 110 \text{ lb} = 1650 \text{ ft-lb}$$

Thus, the amount of work that the robot manipulator generates is 1650 foot-pounds. This work will be accomplished through a transfer of energy.

Power of the Manipulator

The **power** developed in the robot manipulator is the work accomplished within a certain time period. Thus, we can think of power as simply the speed at which work is accomplished. The following equation describes power:

$$P = \frac{D \times F}{T} \qquad\qquad (9\text{–}2)$$

where
  P = power, or the rate at which the work is done, in foot-pounds per second
  D = distance the object must be moved, in feet
  F = force applied, in pounds
  T = time required to move the object, in seconds (s)

The use of Equation 9–2 is illustrated in the next example.

Example
A manipulator moves a weight of 110 pounds a distance of 15 feet in 10 seconds. Find the amount of power developed by the robot manipulator.

Solution
Use Equation 9–2, substitute, and solve:

$$P = \frac{D \times F}{T} = \frac{15 \text{ ft} \times 110 \text{ lb}}{10 \text{ s}} = 165 \text{ ft-lb/s}$$

Thus, the power that the manipulator generates is 165 foot-pounds per second.

Equation 9–2 illustrates the relationship between the cycle time for production and the power of the manipulator. For instance, if the cycle time is decreased, then the power—and, hence, the work—that the robot generates must increase. Thus, if the cycle time for lifting the weight is reduced to 5 seconds, then the amount of power required to accomplish the task is 330 foot-pounds per second. So, Equation 9–2 can be used to determine power (and work) for various cycle times of production.

Energy

Energy is the capacity to do work. Energy exists in many forms, such as mechanical energy, chemical energy, heat energy, or electric energy. One of the basic laws of nature is that energy cannot be created or destroyed. Energy can only be changed from one form to another. For example, the energy developed by an electric motor of a robot manipulator is changed to mechanical energy and is transferred to the end effector through various gears and linkages.

There are two classes of energy in the robotic system: potential energy and kinetic energy. **Potential energy** is energy possessed by an object because of its position. In the previous examples, a certain amount of energy was required for lifting the weight 15 feet above the ground. The energy developed for the task was transferred from the robot's arm to the 110-pound weight. The energy stored in the suspended weight is called the potential energy of the weight. That is, the weight is capable of doing 1650 foot-pounds of work as it returns to its original position. The equation for determining potential energy is as follows:

$$PE = W \times H \tag{9–3}$$

where
PE = potential energy, in foot-pounds
W = weight of the part, in pounds
H = distance the object must be moved, in feet

Kinetic energy is the amount of energy developed when a mass or an object is in motion. The equation for determining kinetic energy is as follows:

$$KE = \frac{1}{2} \times M \times V^2 \tag{9–4}$$

where
KE = kinetic energy
M = mass of the object and includes the relationship of gravity on that mass
V = velocity at which the object will fall when released

In Equation 9–4, the mass and the velocity must be known before kinetic energy can be determined. The **mass** of a body can be calculated from the following equation:

$$M = \frac{W}{32.16} \qquad\qquad (9\text{–}5)$$

where

M = mass of the body, in slugs

W = weight of the body, in pounds

32.16 = acceleration due to gravity, in feet per second squared (ft/s²)

The **velocity** at which an object travels can be calculated from the following equation:

$$V = \sqrt{2 \times g \times D} \qquad\qquad (9\text{–}6)$$

where

V = velocity of the moving body, in feet per second squared

g = gravitational pull of the earth, which is equal to the constant 32.16 ft/s²

D = distance moved, in feet

In robotics, Equation 9–5 is useful in calculating the amount of mass that a gripper will have to lift.

Torque

Torque is the turning effort that is required to rotate a mass (or weight) through a radius. Torque indicates that a weight is present and that the weight is located some distance away from the center of the turning point. Figure 9–2 shows an electric motor and a weight connected to the motor's shaft. The weight is 20 inches from the center of the motor and is 50 pounds. The amount of torque developed from the rotation of this weight can be calculated by using the following equation for torque:

$$T = F \times D \qquad\qquad (9\text{–}7)$$

where

T = torque, in pounds per inch or pounds per foot

F = force, which is the weight of the object, in pounds

D = distance of the object from the center of the rotation, in feet

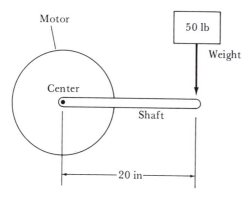

Figure 9–2   Electric Motor and a Weight

The use of Equation 9–7 is illustrated in the next example.

Example
For the arrangement shown in Figure 9–2, find the amount of torque required to rotate the weight.

Solution
Use Equation 9–7, substitute, and solve:

$$T = F \times D = 50 \text{ lb} \times 20 \text{ in} = 100 \text{ lb/in}$$

Thus, the torque developed for turning the weight is 100 pounds per inch. If the weight is changed or the distance from the center is changed, then the amount of torque required to rotate the weight will also change.

Torque can also be expressed per unit of time as it rotates around the center of the motor shaft. This measurement is called the *angular velocity* of the weight. Also, at each of the positions that the weight is rotated, a measurement can be calculated. This measurement is called the *moment* around the center point. The formulas for angular velocity and moment around a center point are beyond the scope of this text. These formulas are used by robotic applications engineers, who must calculate these values.

GEARS

In many cases the actuator that drives the robot manipulator is located at the center of the manipulator. Thus, the energy developed by the actuator must be transmitted from the actuator to the end effector. The transmission of this energy is accomplished through the use of **gears**.

Figure 9–3 illustrates a typical spur gear operation for a robotic manipulator. By definition, **gearing** is the transmission of energy or power through the meshing or contact of the teeth of the gear. In Figure 9–3, the contact point of the teeth of the two gears is the point at which energy is transferred from one gear to another.

The following subsections introduce gear terminology, gear trains, and the various gears found on manipulators, such as the worm gear and the bevel gear. Also discussed is gear adjustment, which is very important to the life of the manipulator's operation.

Terminology

The various gears found on manipulators come in different shapes and sizes. But all of these gears have some common parts. So, in this subsection, we introduce the terminology common to all gears. In the following subsections, these various terms will be used to describe the gears and their operation in a manipulator.

The two gears in Figure 9–3 have special names. The smaller gear is called the **drive gear**, or the **pinion**. The larger gear is called the **driven gear**, the **output gear**, or simply the **gear**. The gearing in Figure 9–3 is a simple **spur gear**. It is called a spur gear because the teeth of the gear stick out like spurs from the base of the gear.

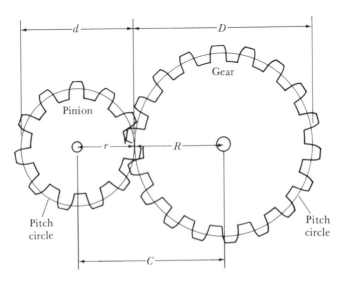

Figure 9–3   Spur Gear Dimensions

Both the pinion gear and the gear are made up of a number of teeth placed at equal distances from each other around the circumference of the gear. The teeth of the gear are cut from this circumference, and the circumference is called the **pitch circle** of the gear. The diameter of this circle is called the **pitch diameter** (represented by D for the gear and by d for the pinion). The radius is called the **pitch radius** (R and r, for the gear and pinion respectively). When two gears are placed as they are in Figure 9–3, the distance between their centers is called the **center distance** (C). When the two gears are meshed, their pitch circles are tangent; that is, they touch each other.

Figure 9–4 illustrates a typical section of a gear and the terms associated with the teeth of the gear. The tooth depth of the gear is made up of two components whose measurements are taken from the pitch circle of the gear. One component is the **addendum**, which is a measurement from the pitch circle to the top of the tooth. The other component is the **dedendum**, which is a measurement from the pitch circle to the bottom of the tooth. The **top land** of the tooth is that part of the tooth that is on top. The **bottom land** of the tooth is that part of the tooth that is in the valleys between the teeth.

Figure 9–4 illustrates another measurement for gear teeth: the **tooth width**. The width of the tooth relates directly to the amount of energy that can be transmitted from one gear to the next.

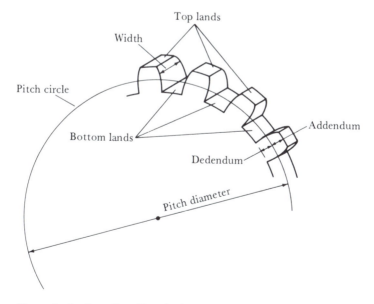

Figure 9–4 Spur Gear Terminology

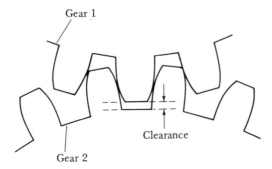

Figure 9–5 Tooth Clearance and the Meshing of the Gear Teeth

The operation of the gear is simple. The addendum of one gear is placed into the dedendum of the other gear. These two gears then mesh to transfer energy from one source to another. A very important measurement is the clearance of the teeth. **Clearance** of the gear teeth is the difference between the length of the addendum of one tooth and the depth of the dedendum of the other tooth, as shown in Figure 9–5.

The **whole depth** of the tooth is the sum of the addendum and the dedendum, as illustrated in Figure 9–6. This measurement does not include the clearance of the tooth. The whole depth of the teeth should always be at least twice the depth of the addendum. This depth will provide a safe operating condition for the gear assembly.

The working depth of the gear is also illustrated in Figure 9–6. The **working depth** of the gear is the depth of the engagement of the two teeth of the gear. This measurement is the sum of the two addendums of the mating gears. The standard working depth is the distance the tooth extends into the tooth space of the mating gear.

Figure 9–6 illustrates two other tooth measurements. The **circular thickness** of the gear is the measure of an arc that is formed

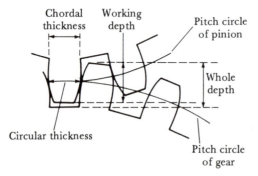

Figure 9–6 Depth and Thickness of the Gear Teeth

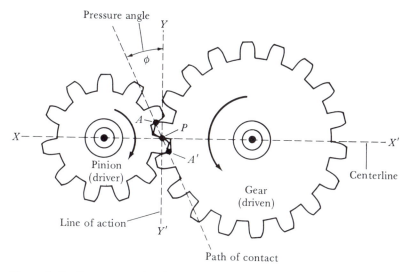

Figure 9–7   Pressure Angle of Meshing Gears

from side to side of the tooth of the gear. The **chordal thickness** is the measure of the chord that is formed by the tooth of the gear.

Figure 9–7 illustrates the **pressure angle** for meshing gears. The centerline, XX', of the gears is a line drawn through the centers of the two gears. When the two gears come into contact, their first contact is made at point A. As the gears rotate, the final contact is at point A'. The path the gears follow from point A to point A' is called the **path of contact**. A **line of action**, YY', which is perpendicular to line XX', passes through point P. Point P is the point at which the two gears mesh. The angle that is formed between the path of contact and the line of action YY' is called the pressure angle, ϕ (Greek letter phi). In the majority of spur gears used in robotics, the pressure angle is 14.5° or 20°.

Figure 9–8 illustrates another important measurement for the gearing operation: the **gear ratio**. The gear ratio is the ratio of the number of teeth on the larger gear to the number of teeth on the smaller gear. For example, in Figure 9–8, the gear ratio is 18:9, or 2:1. This ratio means that the smaller gear will rotate twice for every one revolution of the larger gear.

Figure 9–9A illustrates gear direction for two gears. If the pinion is turning in a clockwise (cw) direction, the driven gear will rotate in a counterclockwise (ccw) direction. Figure 9–9B illustrates an idler gear placed between the driver and the driven gear. The idler gear allows the output gear to rotate in the same direction as the driver gear. Therefore, the driver gear rotates in a clockwise direction, the idler gear rotates in a counterclockwise direction, and, finally, the driven gear rotates in a clockwise direction. The general

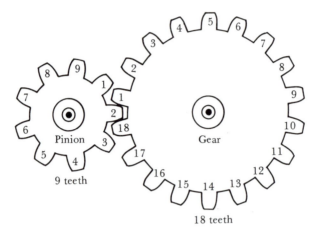

Figure 9–8   2:1 Gear Ratio

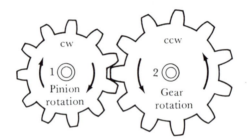

A. Rotational direction for two gears

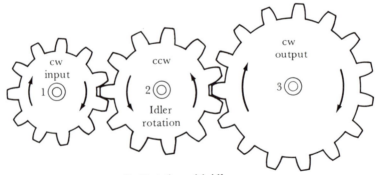

B. Rotation with idler gear

Figure 9–9   Gear Direction

rule is as follows: In a gearing arrangement, if there are an even number of gears, the rotational direction of the output gear will always be the opposite of the rotational direction of the input gear. Likewise, if the number of gears is odd, then the rotation of the output gear will be in the same direction as the input gear.

Gear Trains

A **gear train** is a connection of several gears that produce the correct reduction of force or develop the correct direction of rotation. There are two types of gear trains: ordinary and planetary. An **ordinary gear train** is illustrated in Figure 9–10A. In this gear train, the various sizes of gears are connected in series. That is, the teeth of one gear are connected to the gear next to it. The meshing of the gear teeth causes energy to be transferred from one gear to another.

Figure 9–10B shows a **planetary gear train**. Included in this gearing arrangement is a stationary gear called the **sun gear**. The rotational gear is called the **planet gear**. The planet gear rotates about the sun gear. The motion of the planet gear around the sun gear causes energy to be transferred from the planetary gear train to another gearing system.

The ordinary gear train can be further classified into two types: simple and compound. In the **simple gear train**, each gear has its own shaft. In the **compound gear train**, several gears can be connected to one gear shaft.

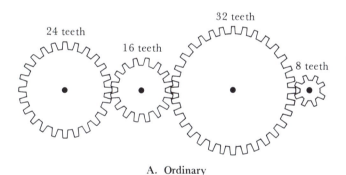

A. Ordinary

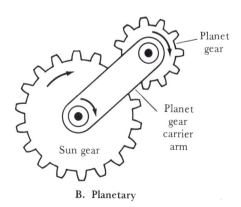

B. Planetary

Figure 9–10 Gear Trains

Figure 9–11 illustrates a typical simple gear train. The formula used to calculate the gear ratio is as follows:

$$\frac{N_D}{N_A} = GR \tag{9-8}$$

where

N_D = number of teeth on the driven gear
N_A = number of teeth on the driver gear
GR = gear ratio of the simple gear train

The gear ratio is for the driver (first) gear and the driven (last) gear. The gears located in between are the idler gears. A simple gear train may be employed to fill the space between two gears or to reverse the direction of rotation between the first and the last gear. Or the gear train's idler gears can be used as power takeoffs for other gears in the operation.

The use of the gear ratio formula is illustrated in the following example.

Example
A simple gear train has five gears. The driven gear has 100 teeth, and the driver gear has 50 teeth. Determine the gear ratio of this gear train.

Solution
We use Equation 9–8, as follows:

$$GR = \frac{N_D}{N_A} = \frac{100}{50} = 2$$

The gear ratio in this gear train is 2 to 1.

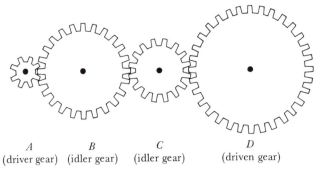

A   B   C   D
(driver gear)   (idler gear)   (idler gear)   (driven gear)

Figure 9–11   Simple, Ordinary Gear Train

As mentioned previously, the compound gear train has more than one gear connected to a single shaft. These gearing ratios will develop the same transfer of energy as the simple gear train does.

A rather complex planetary gearing system is illustrated in Figure 9–12. This planetary gear train is used to establish base rotation of a robot manipulator. The planetary gear train is connected to the hydraulic motor. The rotation of the sun gear causes the planet gear to rotate, thus producing the rotation of the worm gear assembly (worm gears are discussed in the following subsection). The worm gear then drives the pinion gear, which, in turn, drives the boom gear. The boom gear is connected to the rotational base of the manipulator. Since the manipulator's base is heavy, the boom gear must be quite large in order to rotate the base.

Worm Gears

Figure 9–13 illustrates a typical **worm gear** used in manipulator drive systems. A worm gear has two components: the pinion, which is called the *worm gear,* and the driven gear, which is called the *driven worm.* The worm gear is generally used to convert a linear motion to a circular motion. The terms used to describe the worm gear are as follows:

— **Axial pitch,** p_x: The distance between corresponding points of adjacent threads parallel to the axis of rotation.
— **Lead,** L: The distance that a point on the thread will advance for one revolution of the worm.
— **Lead angle,** λ: The angle between the tooth of the worm and the plane of rotation.

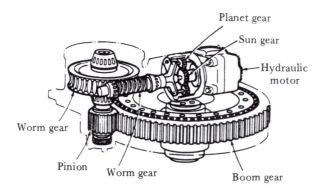

Figure 9–12 Complex Planetary Gear for Providing the Base Rotation of a Manipulator

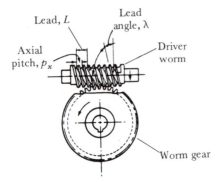

Figure 9–13   Worm Gear

Ball Screw

The **ball screw** is a device used in a gearing operation to convert rotational motion to linear motion and to convert linear motion to rotational motion. The screw assembly is made up of several components, as shown in Figure 9–14. The *input shaft* of the ball screw is the connection of the actuator to the ball screw. The *block* is the connection point of the load to the gearing system. The *screw* is the part of the drive system that is connected to the input shaft and causes motion of the block. In Figure 9–14, notice that the input shaft is rotating in a clockwise direction, while the block is moving toward the input shaft. The *fixed guide shaft* keeps the block moving smoothly across the screw. If the guide shaft were not present, the block would tend to wander and would not move on the screw. The ease with which the screw rotates depends on the two *screw,* or *ball, bearings* connected at each end of the screw. The ball bearings allow the screw to rotate and reduce the friction of movement.

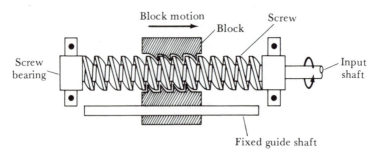

Figure 9–14   Components of a Ball Screw Drive System

Figure 9–15 illustrates another type of ball screw that is used in manipulator operation. This ball screw has ball bearings in its grooves. The ball bearings are held in position by the *ball nut*. The bearings allow the screw to turn easily while the ball nut is moving down the shaft. This ball screw has special cuts in the thread of the assembly to hold the bearings in position.

The ball screw can be used for the R axis or the Z axis of the manipulator. The actuator, connected through a spur gear at the end of the screw, will transmit energy to the thread, which, in turn, will drive the block up and down or in and out. The axis attached to this system will then move linearly. For example, Figure 9–16 illustrates a ball screw used to control the linear motion of a manipulator. The manipulator is mounted to a table on a linear track. The input shaft of the ball screw is connected to a pulley, which, in turn, is connected via a belt to the servo motor. As the DC servo motor rotates, the screw turns and moves the table—and, hence, the manipulator mounted to it—in a linear direction.

Bevel Gears

Bevel gears are used in robotic manipulators when energy must be transferred at an angle. That is, a drive motor is connected to a bevel gear, and the energy is transferred to the driven gear at 90°. The bevel gear is an important component when the point of energy is connected at some distance away from the driving point of the gears.

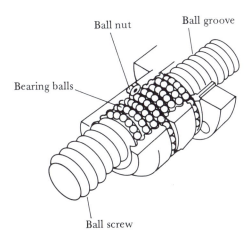

Figure 9–15   Ball Bearings Used with a Ball Screw

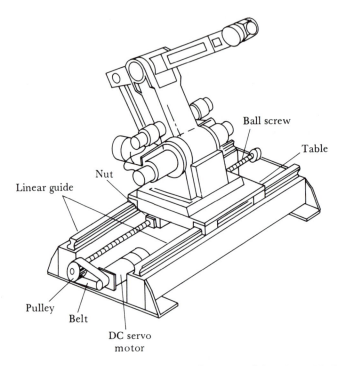

Figure 9–16 Ball Screw Drive Used to Control the Linear Motion of a Manipulator
(Courtesy of GMFanuc Inc.)

Bevel gears are classified into several different types: straight-tooth, zerol, spiral-tooth, and hypoid. In the **straight-tooth bevel gear** (Figure 9–17A), the teeth are cut at a 90° angle to the cone of the gear. The pinion and the gear mesh at a 90° angle. In the **zerol gear** (Figure 9–17B), the teeth are cut on the cone of the gear at angles that form arcs. This cut of the teeth allows the gears to withstand large amounts of load.

The **spiral-tooth bevel gear** (Figure 9–17C) has its teeth cut similar to the way the teeth of the zerol gear are cut. The only

A. Straight-tooth   B. Zerol   C. Spiral-tooth   D. Hypoid

Figure 9–17 Bevel Gears

difference is that this gear is able to withstand greater loads than the zerol gear. In the **hypoid gear** (Figure 9–17D), the cut of the teeth on the cone is much like the cut of the zerol gear teeth. The only difference is that the driver gear is offset from the center contact point of the gear.

The major use of bevel gears in the robotic manipulator is to connect the power from either a 45° angle or a 90° angle. The gear ratio of the bevel gears is 1:1, and the shaft angle of the gear is 90° for robotic applications. This 90° gear connection is generally referred to as a **mitre gear**.

The rotation of a bevel gearing system is illustrated in Figure 9–18. Two output gears are connected to the input gear, which rotates in a counterclockwise direction. Output shaft 1 of the bevel gear rotates in a counterclockwise direction, and output shaft 2 rotates in a clockwise direction. In a robotics application, the input shaft of the gear might be connected to an actuator for the manipulator. As the actuator rotates, input shaft 1 rotates in a counterclockwise direction, and output shaft 2 rotates in a clockwise direction. When these shafts are connected to an end effector, the end effector can develop a two-action move through the use of only one drive actuator.

Gear Adjustment

Backlash is the term used to describe the setting and adjustment of gears. The amount of backlash is the amount of tooth space that exceeds the thickness of the engaging tooth, as depicted in Figure 9–19.

The purpose of the backlash adjustment is to prevent gears from jamming together and making contact at more than one lo-

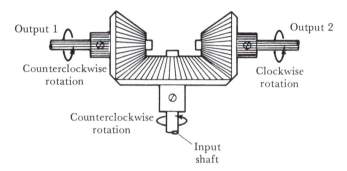

Figure 9–18   Bevel Gear Rotation

Backlash

Figure 9–19 Backlash Measurement for Gears

cation. A lack of backlash results in excessive noise in the gear train, overloading of the gears, overheating of the gears, and even seizing and failure of the gears. Without backlash, the two gears will make contact at more than one point. On the other hand, too much backlash creates problems in gear trains that frequently reverse. If the backlash is too large, then when the gear reverses, the gear teeth will overrun and cause the gear to slip.

Setting the backlash in gear trains depends on the type of gears and the accuracy required of the system. For example, gears that operate at high speeds and that frequently reverse direction require the least amount of backlash. Whatever system is used, the meshing of the gears must be separated enough so that lubricating oil can be placed between the teeth. Without lubricating oil, the heat from the gear will build up. This increase in temperature will cause the gear to expand and thus result in early failure of the gear. Many machinist handbooks on the market have tables that give the correct backlash setting of the various gears and their backlash adjustments.

Some of the concerns in setting the backlash of gears are lubricant space and type of lubricant used, tooth thickness, and thermal expansion of the gear. Some robotic gears might require that the backlash be set at zero. A zero setting means a very tight fit, and for this arrangement, the cost of the gears will increase.

Measuring the amount of backlash is generally accomplished by holding one of the gears stationary and rocking the other gear back and forth. The amount of backlash is then measured with a dial indicator that the technician uses. With spur gears or bevel gears, it does not matter which of the gears is held stationary and which gear is rocked.

Harmonic Drives

The **harmonic drive** is another type of gear found in robotic drive systems. The harmonic drive is made up of three components: the flexspline, the wave generator, and the circular spline. Figure 9–20 illustrates these three components.

Figure 9–20A illustrates the circular spline of the harmonic drive unit. The **circular spline** is a stationary or a rotational component of the drive. The circular spline is a thick-wall component that contains internal spline teeth. The wave generator is illustrated in Figure 9–20B. The **wave generator** is an elliptical, ball bearing component that fits inside the **flexspline** (shown in Figure 9–20C). When the elliptical wave generator is inserted into the nonrigid, cylindrical flexspline, the flexspline becomes elliptical. The elliptical flexspline makes contact with the circular spline at two points that are 180° apart, thus forming positive teeth contacts. These two contact points are shown in Figure 9–20D, which gives the end view of the harmonic drive assembly. The harmonic drive unit is connected to the various actuators of the manipulator to develop the gear reductions and torque required for the manipulator.

Gear reduction from the harmonic drive is created when the wave generator is rotated. One clockwise rotation of the wave generator results in one clockwise rotation of the flexspline. Since the elliptical flexspline makes contact at two points on the circular spline, the gear ratio is 2:1. That is, for every full rotation of the flexspline, the circular spline rotates half a revolution.

Gear ratios for the harmonic drive can be from 80:1 to 320:1. Other harmonic drives offer additional gear reduction ratios. The

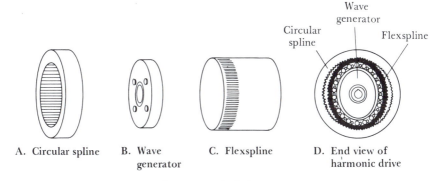

A. Circular spline   B. Wave   C. Flexspline   D. End view of
                           generator   harmonic drive

Figure 9–20 Components of a Harmonic Drive

standard backlash in harmonic drives is zero, and it does not require the backlash adjustments other standard gears require.

BELTS

Because of the physical location of components in many robot manipulators, gears are not the most productive way in which energy can be transferred from the actuator to the load. A better way to transfer energy is through the use of **belts**.

Belts provide a low-cost way to deliver power. Belts also provide a smoothly and quietly running, shock-absorbing system for the robot. Belts are not as strong as gears or chains that might be used for the transmission of power. But belts can be reinforced, so they can be useful parts of the robotic system.

The belts used in the robotic manipulators are V-belts, synchronous belts, and flat belts. Each type is discussed in the following subsections. Belt adjustment is also described.

V-Belt

The **V-belt** is named for its shape. Figure 9–21A illustrates the components of the belt. The belt is made from rubber, which is coated with a jacket that protects the belt from the wear and tear of daily operation. Also, the belt contains reinforcing cords that run the entire length of the belt. These cords give the belt added strength. The cushion section of the V-belt allows the belt to conform to the channel of the pulley on which it may be mounted.

The belt transfers energy through connections of pulleys, or sheaves. These devices allow the belt to turn and transfer its energy to another turning pulley. Figure 9–21B illustrates a belt connected by a pulley. The belt bends around the pulley, but its dimension

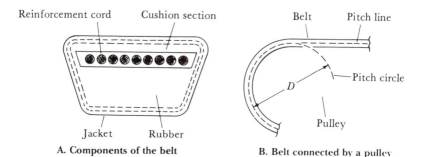

A. Components of the belt

B. Belt connected by a pulley

Figure 9–21 V-Belt

Figure 9–22   Synchronous Belt

will not be distorted, because the size of the belt is taken into account with the pitch diameter of the pulley. The V-belt is used in a manipulator to transfer the energy of the motor to a pulley connected to the base of the manipulator.

Synchronous Belt

Figure 9–22 illustrates a **synchronous belt**. This belt has evenly spaced teeth on the contact surface. It is used with a special type of pulley that has grooves in its surface. The belt's teeth and the pulley's grooves mesh and thus provide correct contact and energy transfer.

Synchronous belts provide the robot's drive system with the positive grip that is needed for the gears or chain drives in the system. The synchronous belt is more expensive than the V-belt, but it provides more positive contact with the pulley. This belt is often referred to as a *timing belt*. The synchronous belt is found in applications where the direction of rotation is constantly changing. For instance, this belt might be used for the wrist assembly of a manipulator.

Flat Belt

Figure 9–23 illustrates the **flat belt**, the most popular type of belt used for transmission of energy in robotics. The flat belt provides

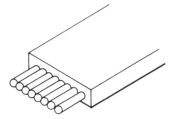

Figure 9–23   Flat Belt

the cheapest method for transferring energy from one source to another at moderate speeds. It also can transfer high amounts of power at low speeds. The flat belt tends to slip on the pulley, and, therefore, it cannot deliver the large amount of torque required in many robot manipulators. The flat belt is generally used when there are short distances between the pulleys and the actuator. For example, the flat belt might be used for the wrist assembly in small manipulators.

Belt Adjustment

All belts used in robotic manipulators tend to stretch and get out of shape. As a result, the amount of energy that is transmitted between the pulleys is greatly reduced. So that correct tension is placed on the belt, adjustments must be made according to the robot's specifications.

The adjustments of the belt can be made either by adjusting the distance of the pulleys or by using the belt adjustor on the belt assembly. Figure 9–24 illustrates a typical method for adjusting belt tension. The tension of the belt is adjusted by pressing the idler wheel against the belt. When the idler wheel is pressed against the belt, the belt becomes tighter. When the idler wheel is removed from the belt, the belt becomes looser. Pressing the idler wheel against the belt takes the slack out of the belt. The depth the idler wheel is pressed into the belt determines the belt's tension. In many cases, the depth of the belt from the horizontal plane will be given by the manufacturer.

CHAINS

In many applications where the slippage of belts is a concern or where energy must be transmitted over a long distance, a **chain** is

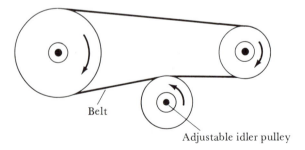
Belt

Adjustable idler pulley

Figure 9–24 Adjusting Belt Tension

the most suitable method of transfer. Chains do not conform to the shape of a pulley by stretching or compressing, as a belt does. And chain drives do not have the shock-absorbing power that belts have. But chains do not vibrate or slip, and they can be used for long-distance transfers of energy.

The most common types of chains used in manipulators are roller chains and bead chains. Each type is discussed in the following subsections.

Roller Chain

The **roller chain** provides high torque transmission and high precision. Figure 9–25 illustrates the typical parts of the roller chain. The chain is made up of many *links*. Each link is comprised of two inside plates, two bushings, two rollers, two pins, and two end plates. The *inside plates* develop the tension for the rollers. The *rollers* make contact with the sprocket and are guided over the sprocket. The *bushings* ensure that the rollers move freely without developing friction. The *pins* hold the bushings and the roller in position. The *end plates* hold the entire assembly together. At the end of the chain, the *master links* hold the complete chain together.

A roller chain and a sprocket are shown in Figure 9–26. The diameter of the roller is important when the chain is meshed through a sprocket gear. That is, the diameter of the roller determines the

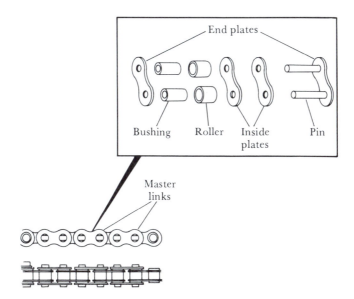

Figure 9–25 Components of a Roller Chain

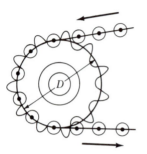

Figure 9–26   Roller Chain and a Sprocket

fit between the roller and the teeth of the sprocket. A poor fit between these two components means that the chain will be sloppy and will not give the correct energy transfer between the actuator and the drive mechanism.

The sprocket gear causes the chain to move along the path. Generally, the sprocket gear is connected to a motor shaft. The rotation of the sprocket provides the transfer of energy from the actuator to the load on the manipulator.

The roller chain is used to transfer energy between the actuator and the gears of the manipulator. Generally, the roller chain is used on long shafts that require a transfer of energy from the actuator to the end effector or the wrist. Figure 9–27 illustrates such an amplification. Also shown in the figure are the devices that are

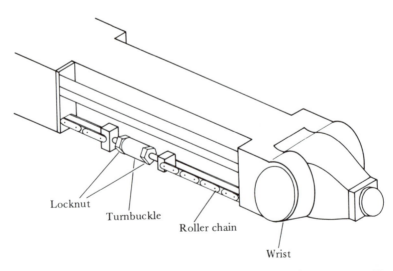

Figure 9–27   Roller Chain for Connecting a Wrist and an Actuator (Courtesy of GMFanuc Inc.)

used to adjust the tension on the chains. The turnbuckle is rotated in a clockwise or a counterclockwise direction in order to make the chain tighter or looser. When the correct amount of tension is obtained, the locknuts are tightened on the turnbuckle to keep it in place.

Bead Chain

The **bead chain** is used for low-torque robotics applications where sensors might be employed. Figure 9–28 illustrates a bead chain. The beads, which are linked together to form the chain, are made of plastic or metal. The bead chain is used with a sprocket that contains dimples. The beads fit into the dimples, causing rotation of the manipulator.

≡ SUMMARY

In many manipulators, the point where the actuator is connected is not the point at which the work is done. So, the energy must be transferred. The energy transfer can be a mechanical transfer, an electrical transfer, a pneumatic transfer, or a hydraulic transfer. This chapter discusses the mechanical transfer of energy, which is accomplished through the use of gears, belts, or chains.

The energy developed through this transfer process allows the manipulator to perform work. This work is accomplished by moving an object through space. The work performed by the manipulator is given by the equation work = distance moved × force exerted. The power of the manipulator is the amount of work that the manipulator can accomplish in a certain period of time. The energy developed by the manipulator is of two types: potential energy and kinetic energy. Potential energy is the energy an object possesses at rest. Kinetic energy is the energy an object possesses when it is in motion.

The torque developed by actuators and manipulators is the amount of energy that is required to turn an object. Torque is de-

Figure 9–28   Bead Chain and Sprocket

termined by two components: the weight of the object and the distance of the object from the center of the rotational point.

Gears are one way that energy can be transferred from the actuator to the manipulator. The energy is transferred through the meshing of the teeth of the gear. Two gears are employed to transfer energy: the pinion, or drive, gear and the output, or driven, gear. The ratio of the number of teeth on the pinion to the number of teeth on the driven gear is the gear ratio.

When two gears are used together, the direction of output rotation depends on the direction of input rotation for the pinion. If the pinion rotates in a clockwise direction, the driven gear will rotate in a counterclockwise direction.

In many situations, the direction of travel or the gear ratio of the gearing system must be changed in order to provide the torque required for the manipulator axis. In this case, the gears can be connected to form a gear train. There are three types of gear trains: the simple gear train, the compound gear train, and the planetary gear train.

The worm gear used in many manipulators converts the linear motion of the axis into a circular motion. The worm gear is comprised of a pinion gear and a driven gear. The ball screw is another type of gear found on manipulators. The ball screw converts the rotation of the actuator into a linear motion.

The bevel gear is used in applications where the motion of the actuator must be directed at a 90° angle. The bevel gear has four different configurations: the straight-tooth bevel gear, the zerol gear, the spiral-tooth bevel gear, and the hypoid gear. Each type is related to the positioning of the pinion gear and the driven gear.

The transfer of energy through the use of gears is accomplished by the meshing of the gear's teeth. The correct amount of meshing is called the backlash adjustment. The backlash of the gear should be set so that the meshing teeth form a good path of contact and pressure angle.

The harmonic drive is another type of gear used with manipulators. The harmonic drive consists of three components: the circular spline, the wave generator, and the flexspline. The wave generator shapes the flexspline into an elliptical shape so that it makes contact with the circular spline in only two places. The ratio of the number of teeth on the circular spline to the number of teeth on the flexspline is the gear ratio of the harmonic drive system.

The transfer of energy from the actuator to the manipulator can also be developed through the use of belts. There are three types of belts used with manipulators: the V-belt, the synchronous belt, and the flat belt. The belt is used only when the transfer of energy is over a short distance. So that the belt makes the correct contact with the actuator and the pulley at the other end of the transmission, the tension on the belt must be adjusted. This adjustment is made according to the manufacturer's specification.

In robotic applications, when energy must be transferred over a long distance, the chain is used. The roller chain is the most popular chain assembly used with manipulators.

KEY TERMS

addendum
axial pitch
backlash
ball screw
bead chain
belt
bevel gear
bottom land
center distance
chain
chordal thickness
circular spline
circular thickness
clearance
compound gear train
dedendum
drive gear
driven gear
dynamics
energy
flat belt
flexspline
gearing
gear ratio
gears
gear train
harmonic drive
hypoid gear
kinematics
kinetic energy
kinetics
lead
lead angle
line of action
mass

mechanical transfer of energy
mechanics
mitre gear
ordinary gear train
output gear
path of contact
pinion gear
pitch circle
pitch diameter
pitch radius
planetary gear train
planet gear
potential energy
power
pressure angle
roller chain
simple gear train
spiral-tooth bevel gear
spur gear
statics
straight-tooth bevel gear
sun gear
synchronous belt
tooth width
top land
torque
V-belt
velocity
wave generator
whole depth
work
working depth
worm gear
zerol gear

QUESTIONS

1. Which area of mechanics deals with the study of the forces that apply to a body at rest?

2. Define the term *kinetics*.

3. Which type of transfer of energy deals with pulleys, belts, and chains?

4. A manipulator produces 2250 foot-pounds of work over a distance of 22.5 feet. What is the force required to develop this work?

5. Define the term *power*.

6. Name two types of energy for a manipulator.

7. What type of energy is developed by a mass in motion?

8. The gripper of the manipulator is holding a part 3 feet above the worktable. The part weighs 10 pounds. What is the amount of mass that the gripper will have to lift?

9. What amount of torque is required to move a 10-pound weight that is located 15 inches from the center of the axis of a manipulator?

10. Give another name for the drive gear.

11. Which measurement of the gear is taken from the pitch circle to the top of the tooth of the gear?

12. On a spur gear, what is a good rule of thumb for the whole depth of the gear?

13. What is the name of the angle formed when two gears mesh?

14. What is the typical range of pressure angles for spur gears?

15. What type of gear is placed between the pinion and the driven gear to cause the driven gear to rotate in the same direction as the pinion?

16. What type of gear train is comprised of several gears connected to one gear shaft?

17. In what type of gear train does a planet gear rotate around a stationary sun gear?

18. What type of gearing system converts a rotational motion into a linear motion?

19. What type of gearing system transfers energy at 90° between the gears?

20. What is the major difference between the straight-tooth bevel gear and the zerol gear?

21. What type of gearing requires the least amount of backlash adjustment?

22. Which component in the harmonic drive causes the flex-spline to become elliptical?

23. What type of linkage is best for a transfer of energy over a short distance between the actuator and the gears?

24. What type of belt can transfer high amounts of power at low speeds?

25. What is the most common type of chain found in manipulator drive systems?

26. Which component on a chain assembly is used to adjust the tension of the chain?

10

Interfacing

OBJECTIVES

Upon completing this chapter, you should be familiar with:

— Interfacing circuitry,
— Interfacing for remote operation,
— Interfacing for the controller,
— Program control of interfacing,
— Connections for interfacing,
— Controller servicing of the input/output lines.

INTRODUCTION

Communication is the key term when one is discussing interfacing operations, that is, the communication between the robotic system and its peripheral components. **Interfacing** allows the robotic controller to communicate with peripheral components located in the work cell of the robot or to communicate with computers in the factory.

Electrical interfacing is accomplished through digital waveforms that are developed within or outside the controller. These digital signals identify the program, the positional information, and the commands for the robot's operation. The communication links of interfacing are either input signals to the controller or output signals from the controller. These signals are standard digital signals that operate on 24-volt DC lines.

Another type of interfacing for the robotic system is mechanical interfacing. Mechanical interfacing is the connection of the end effector to the wrist flange of the manipulator or the connection of the end-of-arm tooling to the wrist flange. These connections are generally made with bolts.

In robotics, there are three basic types of interfacing: electrical interfacing between the controller, the robot, and the peripheral devices; program control of interfacing; and the physical connections for interfacing. This chapter discusses each type of interfacing.

In addition, the chapter describes interfacing for remote control of the robot, since a programmer must often control several robotic work cells from one remote operator's station.

INTERFACING CIRCUITRY

Interfacing is the term used in robotics to describe a communication link between the robot's controller and its peripheral components. These communication links are devices such as programmable controllers, safety fences, or indexing tables. Each communication link gives the controller information it needs for the normal operation of the robotic system. This section describes a simple interfacing link, the output signal circuitry, and the input signal circuitry.

Simple Interfacing Link

Figure 10–1 illustrates a simple interfacing link between the robot controller and the peripheral components within the work envelope

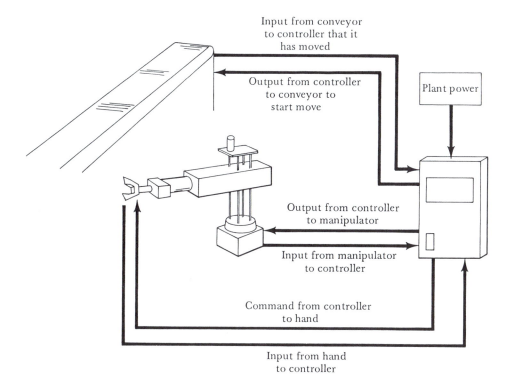

Figure 10–1   Simple Interfacing Link

of the robotic system. The interfacing establishes two types of signals: input signals and output signals.

Signals sent from the controller to the peripheral components are **output signals**. These signals are either on or off signals, and they operate at the DC logic level of the controller, generally, 24 volts. When these signals are on, they are at the 24-volt level. When they are off, they are at the 0-volt level. Output signals are communicated through an interfacing printed circuit board that is located in the controller.

Signals sent from the peripheral components to the controller are **input signals**. These signals are generally at the same $+24$-volt DC level as the output signals. Also, the input signals are processed through a different printed circuit board from the output signals. The logic levels associated with these signals are on and off. Input signals are connected to an external, to the controller, $+24$-volt DC power supply.

Another communication link is established between the robot controller and the end effector of the manipulator. The end effector requires special signals that tell the end effector's actuators to activate or not activate. For example, in the interfacing operation shown in Figure 10–2, the signal is developed from the controller to the end effector. The output signals from the controller tell the end effector to open its grippers. An output signal of $+24$ volts DC is supplied to the pneumatic actuator on the end effector. The pneumatic actuator activates, and the grippers open. Removing the $+24$-volt signal from the controller tells the gripper to close.

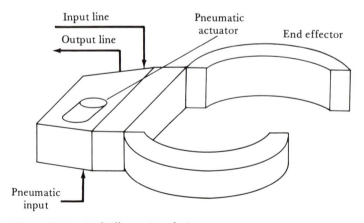

Figure 10–2 End Effector Interfacing

Output Signal Circuitry

The controller contains an interfacing board with two sets of circuits: the output, or transmitter, circuits and the receiver circuits. In this section, we will focus on the output circuits.

Figure 10–3 illustrates several transmitter circuits. In the figure, the interfacing signal is developed from the controller and processed to the base of a transistor. This signal from the controller turns on the transistor and causes conduction through the emitter-collector circuit of the transistor. When the base voltage is lowered, the transistor will turn off. The transistor in this circuit acts as a switch. Notice that the transistor operation depends on a +24-volt DC power supply that is not part of the controller. This power source is provided by the robot installer. The main reason for this is to ensure that plant-level power fluctuations do not interfere with the computer control of the controller.

The output of the transistor is fed to some type of control circuitry. The control circuitry might be components to indicate that an output signal is being sent from the controller. For instance, the user of the robotic system might want to place a relay or a light in the output circuit to indicate that there is an output condition.

Figure 10–4A illustrates a connection of a relay to the output circuitry. The coil of the relay is connected to the collector of the transistor. When the transistor is turned on, it will conduct and cause current to flow through the relay's coils. The current through

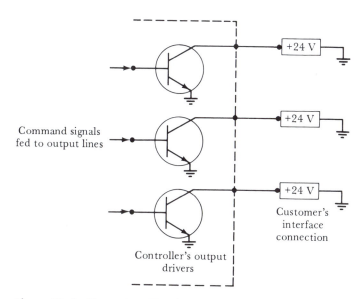

Figure 10–3 Transmitter Circuits

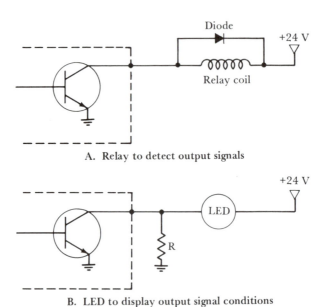

A. Relay to detect output signals

B. LED to display output signal conditions

Figure 10–4   Connection of Components to Detect the Controller's Output Signals

the coil will close the relay's contacts. In an actual application, the contacts of the relay might be connected to the motor circuitry of an indexing table. Their closing will cause the table to advance to the next location for robot operation. Notice that the coil of the relay is connected to a diode. This diode protects the circuit from reverse current created by the coil. Reverse current would cause the transistor to short out each time the relay operates.

Figure 10–4B illustrates another method of detecting an output signal. In this example, a light-emitting diode (LED) is connected to the transistor collector. When the transistor turns on, current flows through the LED, thus turning it on. When the LED is lit, it should be at only half brilliance. The resistor connected in parallel with the collector-emitter of the transistor causes the lamp to draw only half power. Again, this connection protects the output transistor.

The signal developed at the output may be either a pulsed output or a continuous output. Figure 10–5A illustrates a pulsed waveform. Here, the signal is pulsed for at least 200 milliseconds, giving the output enough time to read the signal and respond. In other words, if the output signal is applied to the coil of the relay, then the pulsed signal must remain on for at least 200 milliseconds in order for the coil to react.

Figure 10–5B illustrates a continuous signal. This signal will remain on as long as power is applied to the controller's signal. As

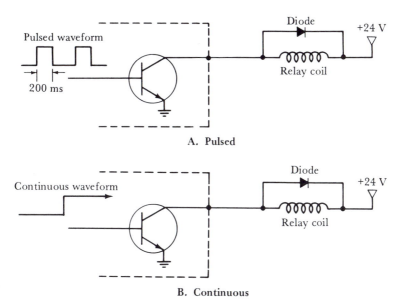

Figure 10–5 Signal Output Conditions

an example, continuous output might be used to turn on a lamp indicating that the robot is in operation, providing a safety feature for the robotic system.

Input Signal Circuitry

The input signal can be developed from many different peripheral devices or from the end effector all sending signals back to the controller. Figure 10–6 illustrates the interfacing connection of the input signal.

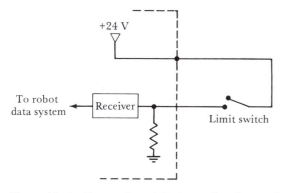

Figure 10–6 Connection of the Input Signals to a Controller

The schematic shows the various components that are used to interface the external signal to the controller. The components are an internal 24-volt power supply in the controller, an external limit switch on the peripheral device, a receiver circuit located in the controller, and a current protection resistor. The number of input circuits on the controller can vary depending on the technology level of the controller. The low-technology controller may have only 3 or 4 input lines. The medium-technology controller may have 10 or 12 input lines. The high-technology controller may have 32 input lines that can be connected to peripheral devices. In new controllers, input modules are designed to be adapted to the user. This means the user could select from a minimum of 16 inputs to a maximum of 132 inputs.

The input lines on the controller are used to sense the closing or opening of limit switches. The limit switch on the peripheral device is used, for example, to sense whether a part is present. If a part is present, the limit switch will close, and the controller will sense voltage at the receiver input. The controller's microprocessor also scans the input lines for a switch closing or opening. When a switch closure is detected by the controller's microprocessor, the robot program will develop an operation based on this input.

Another example of this operation involves a safety fence. The limit switch is connected to a safety fence located around the periphery of the manipulator's work envelope. A gate in the fence provides programmer access to the area. The gate also has a limit switch, which, in turn, is connected to an input line on the controller. When an individual opens the gate, the limit switch is closed, which, in turn, generates an input signal on the input interfacing line. This input is detected by the microprocessor, which tells the manipulator to stop motion.

The input circuitry looks for an input signal that develops a voltage in the range of 20–28 volts DC. This voltage range will develop a logical 1 state, indicating that information is on the line. If the input voltage varies from 0 to +4 volts DC, then the line is considered to be logically low, and no information is present.

Each of the receiver circuits on the controller develops a certain amount of input resistance. The typical resistance value is 3.3 kilohms, which is the value necessary for correct operation of the receiver circuit. Large resistance in the receiver circuit ensures that when the signal goes high, it will stay in the 20–28-volt DC range. High resistance also ensures that when the signal goes low, the voltage range will be 0–4 volts DC. Resistance levels below 3.3 kilohms may cause false readings at the receiver circuitry. For instance, the receiver might show a low input when it should show a high input. If the program in the controller is set up to perform

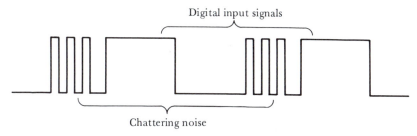

Digital input signals

Chattering noise

Figure 10–7 Chattering Developed on the Input Lines

an operation based on a high signal from the receiver circuit, the program will not run, because a high input signal is not read by the controller.

The signal of the input must be on for a short period of time so that the line can be read. In general, the input time of the cycle is about 200 milliseconds. Any signal of shorter duration might not be read by the controller.

If relays are used as input devices, the designer of the input circuitry must take their chattering noise into account. **Chattering noise,** which develops from the closing of the limit switch, is illustrated in Figure 10–7. If chattering time is long, the input might read chatter as a high condition. As a result, false information would be fed to the controller. Typically, chattering time must be lower than 5 milliseconds.

When a closed circuit is being read, the resistance of the input circuit must be low. Generally, the resistance of the input component should be no greater than 100 ohms. When the input circuitry is off, the resistance of the component should be at least 100 kilohms. These resistances ensure that good signals are developed when the switches are inputting data and that no false information is transferred when the switches are open.

INTERFACING FOR REMOTE OPERATION

In many applications, the controller is located close to the manipulator, while the master controls for operating the robot are located at a remote station. Figure 10–8 illustrates the connection of two work cells to a remote operator's panel. The operator of the work cell is located at the operator's workstation. Here, the operator can monitor various signals that are sent from the robot controller to the operator's workstation. These signals are connected to the remote interfacing connector on the robot controller and to the remote signal connector on the operator's panel.

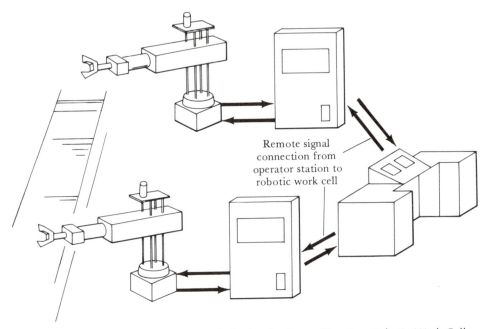

Figure 10–8   Remote Operator's Station for Controlling Two Robotic Work Cells

The interfacing between the controller and the operator's workstation is of two types: the connection of the cables between the operator's station and the controllers, and the signals that command the robot's operation from the workstation. Since the operator must monitor the operations of several work cells simultaneously, the operator needs to see certain signals from the controller on the remote operator's panel. These signals can be connected through the interfacing operation to the remote location.

The following subsections describe the signals that are inputted to and outputted from the remote operator's panel and the remote program selection that is available to the operator.

Remote Operator's Panel

Since the operator will not be near the robot controller, the operator must use signals to command the robot to perform certain tasks. Figure 10–9 illustrates the various signals that might be found on the remote operator's panel. The signals are input and output signals.

The output signals are connected to the controller. When these signals are inputted to the controller, the controller commands the robot to perform a certain operation. One such output signal is the start line. The start command from the operator's panel interfaces

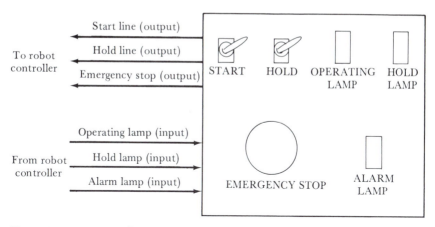

To robot controller

Start line (output)

Hold line (output)

Emergency stop (output)

From robot controller

Operating lamp (input)

Hold lamp (input)

Alarm lamp (input)

START   HOLD   OPERATING   HOLD
                 LAMP   LAMP

EMERGENCY STOP   ALARM
                  LAMP

Figure 10–9 Input and Output Signals on a Remote Operator's Panel

with the controller and causes the robot to start an operation. Generally, the robot has a program called start that causes the robot manipulator to calibrate.

The hold line is another signal connected from the operator's panel to the controller. The hold signal causes the robot to stop all motion and remain at the position it was in when the hold signal was given. The final output signal shown on the operator's panel in Figure 10–9 is the emergency stop signal. Again, this signal is connected from the operator's panel to the controller. This signal causes the robot to stop at once.

Indicator lamps (input lines) are also found on the remote operator's panel. These signals are sent from the controller to the operator's workstation. The input lines cause a lamp to illuminate on the operator's panel, telling the operator that the robot is in cycle or that an alarm condition has developed in the system. When the operating lamp is on, it indicates that the robot is in normal operation. The line for the hold lamp becomes high when the robot has been placed in the hold mode. The alarm line becomes high whenever an alarm condition is developed in the robotic system. For example, this line will go high when the emergency stop button is depressed on the operator's panel.

Figure 10–9 also illustrates three switches that may be found on the remote operator's panel. Generally, the safety of the worker and of the equipment is the main concern at a remote workstation. Thus, the switches on the operator's panel relate to safety features. The switches on the panel are normally open switches. That is, the switch must be closed by the operator before the signal can be sent. Also, if the switch develops a problem, the switch will remain open and not send an incorrect signal to the controller. The three safety switches on the operator's panel shown in Figure 10–9 are the start switch, hold switch, and emergency stop button. These three switches are the most common switches on a remote operator's panel.

Remote Program Selection

Calling up programs from a remote location is a basic part of an automated robotic work cell. The operator located at a remote work-station can call up various programs. The robot processes these signals through a **remote program selection function**. This circuitry allows the operator to select, through switch settings, the various programs that are stored in the controller. These stored programs are coded in such a way that they can be called up through the binary input established by the setting of the switches on the remote operator's panel. In this operation, the remote operator's panel is interfaced with the robot controller. The lines connected to the re-mote panel provide the controller with the necessary 24 volts DC, which indicate a high condition.

The selection process is illustrated in Figure 10–10. The input interface is tied to four lines, and each line is tied to the +24-volt

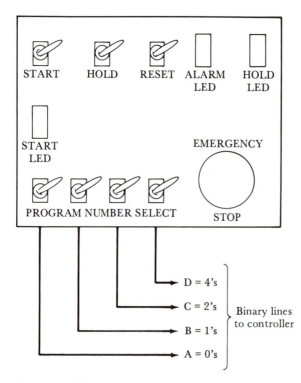

Figure 10–10   Binary Output from a Remote Operator's Panel

DC power supply required for the input interface. Each of the contactors has a weighted bit in binary. For example, switch A has a binary bit of 0; switch B = 1, switch C = 2, and switch D = 4. The arrangement of these various switches will develop numbers from 0 to 7. Thus, if switches C and B are closed, then the coded binary number is 3. This coded input is read by the controller, and the controller then calls up the program that has been stored in section 3 of the remote program call memory.

INTERFACING FOR THE CONTROLLER

The controller has several interfacing ports for connection with various external peripheral components. Three interfacing connections are discussed in this section: computer interfacing, external memory interfacing, and sensor interfacing.

Computer Interfacing

Input ports are provided on the controller for interfacing the various computer controls. These input ports are generally either RS–232 or RS–422 format. The input information is fed to the interfacing connection in a *serial format*. That is, the information has a start bit, data information, a parity bit, and a stop bit, as we showed earlier in Chapter 3. The information sent over the computer interface is called **full-duplex**. This term means that the port for the computer interface can send data and receive data over the same line.

   With the computer interface, the programmer of the robotic system can program off-line. That is, all of the service codes, geometric moves, and velocity rates can be programmed off-line and then downloaded to the controller.

   Sending data via the RS–232 format can only be done for about 50 feet. After that, the data will start to lose its amplitude and pick up stray noise on the line. This stray noise will show up in the data transmission as false information to the controller. The signal level can be kept constant over the transmission length of the cable through the use of line amplifiers, which are nothing more than signal amplifiers that maintain or increase the signal level.

External Memory Interfacing

Many controllers for medium- and high-technology robots have some type of external memory capacity. Thus, a program that has been operating in the controller can be downloaded to an external memory device and used in a later operation. The external memory device might be a bubble memory, a magnetic tape, or a floppy disk.

The controller must have an interfacing port for the external memory device. Figure 10–11 illustrates the connection of this port. The format used for this data transfer is RS–232.

Figure 10–12 illustrates another type of external memory device that interfaces with the controller. This memory device uses bubble memory for the storage of program information. In the figure, the bubble memory cassette is connected to the controller's door. The operator places the bubble memory into the holder, which interfaces the cassette to the controller. The operator then commands programs to be downloaded from the controller to the memory for storage or to be uploaded from the memory to the controller for robot operation.

Sensor Interfacing

One of the most important, recent applications in robotics is the sensor application. The sensor can communicate touch, vision, or proximity information to the controller. This information is converted to digital signals before being sent to the controller. An input interfacing port on the controller provides for the connection of the sensor. This data input port is generally an RS–232 port with full-duplex capabilities.

Information sent through this port is placed on the data bus and sent to the controller's internal memory. The data on this line is generally under program control. Thus, special codes must be used in the program for reading this sensor data input.

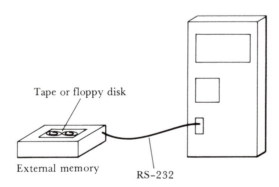

Tape or floppy disk

External memory

RS–232

Figure 10–11 Interfacing Port for External Memory

Figure 10–12 Operator Loading a Bubble Cassette Memory (Courtesy of GMFanuc Inc.)

PROGRAM CONTROL OF INTERFACING

Program control of interfacing is broken down into two classifications: control over the peripheral components, called **service requests**, and control over the robot's end effector, or **robot requests**. The following subsections discuss these two types of requests.

Service Requests

The interfacing operation for peripheral components provides program control of the periods when data are transmitted and received by the interfacing ports. Program control of the sequence of events is critical to the successful operation of the robotic cell.

Figure 10–13 illustrates several input and output signals that must be interfaced by the programmer. In the figure, the robot is operating within a work cell, and the controller is responsible for the total operation of the work cell. To control the sequence of events, the programmer must consider when the signals will be outputted from the controller and what types of signals will be needed by the work cell.

The programmer might first consider what signals will be required. Figure 10–13 indicates that several output signals are required. The programmer might list all the output signals, that is, the commands from the controller, as follows:

— Command to advance the indexing table;
— Command to open the lathe door;
— Command to activate the proximity sensor;
— Command to close the lathe door;
— Command to start the machine cycle;
— Command to open the lathe door after the machine cycle has been completed;
— Command to release the part from the lathe chuck;
— Command to advance an empty pallet on the indexing table for deposit of a finished part;

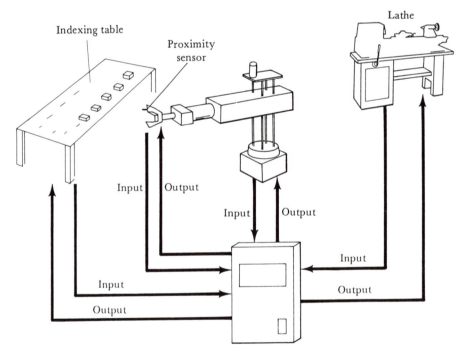

Figure 10–13 Input and Output Signals to Be Interfaced by a Programmer

— Command to advance another part on the indexing table in front of the manipulator to start the cycle over again.

Input signals to the controller are also required to ensure that the robot's tasks have been completed. Thus, the programmer might make a list of the various input signals required by the controller, as follows:

— Signal that the lathe door has opened;
— Signal that the machine is in the start cycle;
— Signal that the proximity sensor has sent back data from the chuck;
— Signal that the indexing table has advanced;
— Signal that the machine has started its cycle.

The programmer might now develop a flowchart of the program and specify when each of these input and output signals must happen.

Robot Requests

Robot requests also require input and output signals. And as indicated in Figure 10–13, the controller is responsible for control of the end effector.

Once again, the programmer should make a list of the various control signals for the tasks of the end effector. These control signals might include a signal to open the grippers, a signal to close the grippers, and a signal to monitor whether a part has been picked up by the hand.

There are only two output signals required for the operation of the end effector in this example. They are as follows:

— Open gripper,
— Close gripper.

The input signals tell the controller that the different tasks have been completed. For this example, the input signals required by the controller are as follows:

— Part present in gripper,
— Hand broken,
— Hand closed,
— Hand opened.

The input and output control signals tie the entire robotic

system together. The interfacing provides communication between the different components so that the work cell functions.

CONNECTIONS FOR INTERFACING

The interfacing connector must be attached to the controller. In many cases, the connector will have to be hard-wired to the controller. But the connectors for interfacing the external memory port or the RS–232 port for the sensor, for example, are provided by the manufacturer of the controller.

Figure 10–14 illustrates a typical connector for robot interfacings. This connector is an input/output service connector. That is, this connector can be attached to the peripheral components. Notice that the connector has a pin for each of the output connections, with a total of 50 pins available. The connector supplies both the input and the output service request signals. The connector also provides the 24-volt DC source to the peripheral components, and it provides some signals for the remote operator's station.

1	RSR1			33	SDO1
2	RSR2			34	SDO2
3	RSR3	19	START	35	SDO3
4	RSR4	20	RDY	36	SDO4
5	AUTO	21	AUT	37	SDO5
6	*SFSD	22		38	SDO6
7	*ESP	23	ALM	39	SDO7
8	*HOLD	24	SDI9	40	SDO8
9	SDI1	25	SDI10	41	SDO9
10	SDI2	26	SDI11	42	SDO10
11	SDI3	27	SDI12	43	SDO11
12	SDI4	28	SDI13	44	SDO12
13	SDI5	29	SDI14	45	SDO13
14	SDI6	30	SDI15	46	SDO14
15	SDI7	31	0V	47	SDO15
16	SDI8	32	+24V	48	SDO16
17	0V			49	0V
18	+24V			50	0V

Figure 10–14   50-Pin Connector for Interfacing Components

CONTROLLER SERVICING OF THE INPUT/OUTPUT LINES

The microprocessor that operates the controller is responsible for servicing the various input and output interfacing lines. When data from the controller must be sent over the interfacing lines, the microprocessor sends the data to a temporary storage device. This temporary storage device is called a *register*. The data is stored in the register until the robot or the peripheral device is ready for the data.

The input of data for the interfacing operation is more complex. The input data is received from the manipulator's end effector and also from the peripheral devices. When data is inputted from an external device to the controller, the microprocessor uses a special routine called polling. In **polling**, the microprocessor scans the input and output interfacing lines and determines which line requires service.

The microprocessor also has a special routine that allows it to wait until the input or output interface is ready to transmit data. For example, suppose that the controller must receive program data from an external memory source in order for the program to operate. In this situation, the microprocessor will wait until the interfacing port is ready to transmit data. If the microprocessor does not have any request for service on the interfacing lines, it will continue to operate the program.

In case of emergency, the microprocessor has a routine that interrupts the program that is running. For example, suppose that a person enters the work envelope of the robot. For safety of the person and the robot, the robot should stop its work. This interruption can be accomplished through the use of a limit switch connected to the alarm input circuitry. When the switch is turned on, a signal is sent to the interfacing input port. The microprocessor recognizes this situation as an emergency and stops normal operation. This external interrupt stops the program that is running in the microprocessor and stops the robot's movements.

SUMMARY

Interfacing is the communication between various components of the robotic work cell. The communication link is established between the peripheral devices, the controller, and the manipulator's end effector.

Two types of signals are involved in the interfacing of the robot controller: input signals and output signals. The input signals are developed through a receiver circuit within the controller; the output signals are connected through transistors.

The input and output signals are called digital signals because the signals are either high or low. The high digital signal operates from a +24-volt DC source; the low digital signal operates from a 0-volt DC reference. The times at which these signals are either on or off are also controlled by the robot's controller.

The interfacing operation requires the programmer of the robot to assign various lines as either input lines or output lines. If the programmer wants special lines, then these lines must be connected so that the interfacing operation can be controlled through programming methods.

The interfacing operation may be developed through a remote connection. This process allows the operator of the workstation to control different work cells from a central location. Status signals from the robot can be sent to the remote workstation, and signals from the remote workstation can be sent to the robot controller. These signals are wired through cables that connect the two locations.

In many high-technology automated systems, computers are interfaced with the controller. The programs are written on the computer and then downloaded to the robot controller. The information is sent by using an RS–232 or an RS–422 format for the data. External memory is another type of interfacing for the robot controller. External memory may be a magnetic tape that is connected to the controller or a bubble memory cassette. These external memory devices allow the user to download and upload data for the operation of the robot.

The robotic work cell is meant to be automatic. In other words, a program can be stored in the controller and later recalled by the programmer. The data can be called up through a remote location interfacing with the controller, such as a peripheral device that requests service through the controller. Signals that are sent from the controller to peripheral devices are called service requests. Signals sent from the peripheral device to the controller are called robot requests.

The main controlling device within the robot is the microprocessor. This component scans the various interfacing lines and reads the data on these lines. If a digital input line contains data, the microprocessor uses the data and performs the necessary operation to complete the robotic task.

KEY TERMS

chattering noise   polling
full-duplex port   remote program selection
input signals   function
interfacing   robot request
output signals   service request

QUESTIONS

1. Describe how communication lines are established between the controller and the peripheral devices.

2. What type of signal is sent from the controller to the peripheral device?

3. Name two robotic system components to which an output signal can be connected.

4. What commands the gripper to open and close?

5. What causes the conduction of current through the emitter-collector of the output transistor?

6. Where is the power supply located for the output circuitry for interfacing signals?

7. What type of device can be placed in the output circuitry to give an indication that an output signal is present?

8. Why is a diode connected across the coil of the relay?

9. How can the power developed through the collector-emitter of the output transistor be cut in half?

10. Name two types of output signals for the interfacing operation.

11. Give an example of the use of a continuous output signal.

12. What logic level indicates that data is present on the line?

13. What is the logic voltage level for an input signal?

14. What happens when the input signal is less than the 200-millisecond length?

15. What is the main reason for using remote operator workstations?

16. Name three input signals that may be found on the remote operator's panel.

17. What type of computer port on the controller is able to transmit data as well as receive data?

18. What information relating to the robot program can be done in an off-line mode?

19. Name three types of external memory devices.

20. What type of port is used for sensor interfacing?

21. What is the name of the input signal that is sent from the peripheral device to the controller?

22. What concerns must the programmer be aware of when establishing a programming flowchart for interfacing operations?

23. Where are robot requests developed?

24. In the following list, classify the signals as either a robot request or a service request: (a) part present in gripper; (b) lathe door opened; (c) hand closed; (d) index table; (e) weld.

25. Name three signals found on the interfacing connector on the controller.

11

End Effectors

OBJECTIVES

Upon completing this chapter, you should be familiar with:

— Basic considerations for end effectors,
— Gripper design,
— Mechanical grippers,
— Vacuum grippers,
— Magnetic grippers,
— End-of-arm tooling.

INTRODUCTION

The purpose of the robot manipulator is to perform work. The work performed by the robot manipulator must be accomplished by an **end effector** attached to the end of the robot's arm. The end effector can be a gripper (hand) or end-of-arm tooling.

The manipulator is responsible for moving the end effector to programmed locations. These moves of the end effector are controlled through the robot's program. Once the robot has moved the end effector to its location, the particular task that is programmed for the end effector will be performed.

Many different designs can be used as an end effector, depending on the task to be performed by the robot. The robot's end effector can be designed to mechanically grip a part, to use a vacuum to lift and transfer a part, or to use an electromagnet to lift and move a part.

The design of the end effector is very important. That is, the end effector must be flexible enough to be retrofitted to perform another task, without much redesign, once a job has been completed.

In this chapter, the various end effectors used in the robotic industry will be examined. The design, lifting characteristics, and applications of each of the end effectors are also discussed.

BASIC CONSIDERATIONS
FOR END EFFECTORS

The manipulator must be able to move parts, weld, apply sealant, or assemble components. For performing these tasks, the manipulator has an end effector, which may be a gripper or an end-of-arm tooling. The end effector of the manipulator will cause certain problems that the programmer must be aware of. These problems involve program control, the size of the work envelope, cycle times, and the safety of the worker. In the following subsections, these problem areas will be explored. First, though, we will examine the types of end effectors.

Types

End effectors are divided into two classifications: grippers (or hands) and end-of-arm tooling (EOAT). The **gripper** is used to lift parts or to transfer parts from one location to another. A typical gripper is illustrated in Figure 11–1. In this figure, the gripper is responsible for picking up parts from an assembly line and placing the parts into the chuck of a machine. With the same gripper, the robot then picks the part from the chuck and palletizes the parts on the table. The gripper in this application is doing nothing to the part itself.

Figure 11–2 illustrates end-of-arm tooling. **End-of-arm tooling** makes changes in or operates on a part. For example, the end-of-arm tooling could arc-weld, spot-weld, debur, grind, apply sealant, rout, drill, glue, or drive screws. In Figure 11–2, an arc-welding

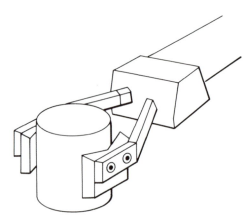

Figure 11–1   Gripper Used to Transfer Parts

Figure 11–2 End-of-Arm Tooling Used for Arc Welding (Courtesy of The DeVilbiss Company)

torch is attached to the end-of-arm tooling. As the robot moves through its programmed paths, the welding torch may, for instance, lay down a weave pattern.

The end effector is generally mounted to the end of the manipulator's arm. This mounting is called an end effector mounting flange. In the majority of industrial robots, a bolt hole pattern is placed on this flange. The holes are arranged so that the end effector can be mounted in several different planes. The end effector is then screwed into the flange for a secure fit.

Each robot manipulator is able to lift a certain number of pounds. This lifting capacity of the manipulator is called its **payload**. For example, a certain robot may be capable of lifting 10 kilograms.

If an end effector weighing 2 kilograms is mounted on the manipulator, then the maximum lifting capacity of the robot is reduced to 8 kilograms. Thus, the user of the robotic system must know the weight of the end effector in order to determine the robot's actual payload. If the weight added to the end of the robot's arm exceeds its specified payload, then the repeatability and accuracy specifications of the robot will suffer.

Program Control

The programmer of the robot can control the action and the path of the end effector through the program. Thus, the programmer is able to move the end effector to the same location over and over again.

To illustrate the use of program control of a gripper, let us consider the application shown in Figure 11–3. The programmer wants the gripper to pick a part from the conveyor line and palletize these parts in three separate piles. The boxes are pushed into location with a push cylinder, and upon command from the robot controller, the gripper reaches into the stack and removes four boxes

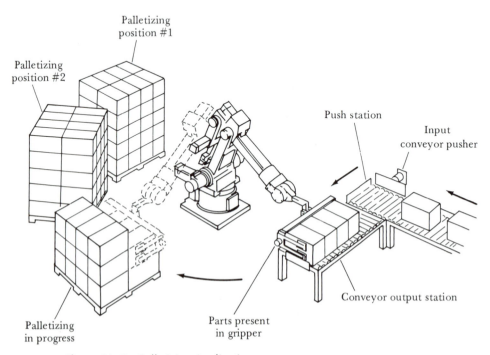

Figure 11–3 Palletizing Application

at a time. Once the gripper is in place around the boxes, a control signal from the controller commands the gripper to squeeze the four boxes. The robot will stay at the pickup point for a certain period of time to ensure that the boxes are in the gripper's grasp. The pressure applied here is only enough to hold the boxes in place. The boxes are then put on the pallets behind the robot.

The programmer will have to make several point-to-point moves to have the robot complete this task. The first move is a linear move in front of the boxes. A second linear move is required when the robot approaches the boxes. The manipulator must make a third linear move away from the conveyor. Then, it must swing around and deposit the boxes on the pallet behind it. The manipulator, after depositing the boxes, returns to a perch position until it is commanded to pick up another set of four boxes. The gripper in this example might have sensors in its fingers to alert the controller that all four boxes have been picked up.

Through program control, the programmer can activate the end effector. The command for the gripper to close is outputted to an actuator solenoid valve on the gripper. The valve activates the closing of the gripper. A similar command from the controller opens the gripper, releasing the boxes.

Work Envelopes

The programmer must be aware that the addition of an end effector to the manipulator will increase the size of the robot's work envelope. For instance, a typical articulate robot arm has a work envelope of about 6 feet. When end-of-arm tooling is added to the manipulator, the size of the work envelope increases by 1.5 feet. This simple addition of 1.5 feet to the size of the robot's work envelope might mean that the robot now will not fit in the area for which it was planned. Also of importance is the size of the part the robot will be moving. If the part extends beyond the gripper, it will also increase the size of the work envelope. Thus, the designer of the robotic work cell must take the size of the end effector and part size into account when planning a robotic application.

Cycle Times

The cycle times of the robot are very important in meeting production deadlines. And the movement of the robot's end effector plays a crucial part in the cycle time. The robot's end effector should be articulated only in small movements to ensure that cycle times are met. Several methods can be employed to ensure that a robot with an end effector will perform its task within the correct cycle times.

The parts presented to the robot should always be placed in front of the manipulator. Placing the part in front of the manipulator means that the robot needs to make only short moves to grasp a part. Thus, the orientation of the workpiece can shorten the cycle time.

In many applications, a gripper that matches the job can shorten the cycle time. Figure 11–4 illustrates a **dual gripper** that can be used in loading and unloading machine parts. Figure 11–4A gives a schematic view, and Figure 11–4B gives a full view. One gripper can pick up a part, and the empty gripper can remove the part from the chuck of the machine. For this type of application, the robot first goes to a pick position and loads an unmachined part into the gripper. The robot then moves into position in front of the machine. When the machine cycle is finished, the robot reaches into the machine and unloads the finished part. The robot manipulator then moves away from the machine cell, and the gripper is flipped 180°. The manipulator then moves the gripper into the lathe so that the unmachined part can be placed into the chuck. The machining cycle starts once the robot has removed the gripper. The finished part is then palletized.

The dual gripper allows the robot to meet the production cycle time. That is, the dual gripper can hold two parts at one time: the finished part and the unfinished part. While the machine is in cycle,

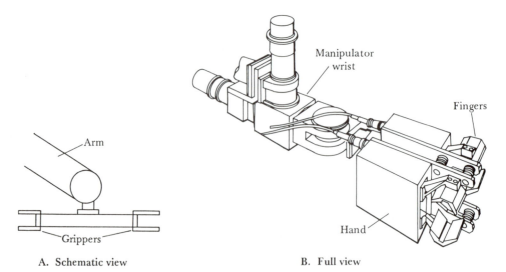

A. Schematic view   B. Full view

Figure 11–4 Dual Gripper

the manipulator can be programmed to perform another task. When the machine cycle is complete, an interfacing signal is outputted from the machine, and the manipulator moves into position to pick up the finished part and load the unfinished part. These moves allow the manipulator to perform the task as a human would.

Safety

Figure 11–4A shows an end effector connected directly to a mounting flange. With this type of connection, if the robot's arm were to move into a position where the end effector would crash into the fixture, damage would result to the end effector, the fixture, and the robot manipulator. So that damage to the robot is prevented, a **safety joint** is attached between the manipulator and the end effector, as shown in Figure 11–5. In this case, if the end effector crashes into a fixture, the safety joint will break, separating the end effector from the manipulator.

In many high- and medium-technology controllers, an electronic circuit is added to the safety joint mounting. Figure 11–5 illustrates an electronic switch added to the end effector's flange. The switch is a normally closed switch. When the safety joint is broken, the switch is activated. The switch then sends a signal to the robot controller, alerting the controller that the safety joint has broken. This alarm condition causes the robot to cease operation and develops an error message to alert the operator.

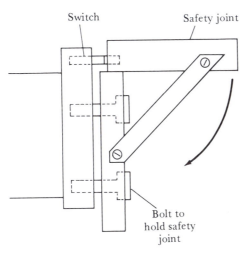

Figure 11–5 Safety Joint and an Alarm Switch for an End Effector

≡ GRIPPER DESIGN

The gripper is composed of two sections: the **fingers** that grip the part and the actuating circuitry that causes the fingers to open and close. In our discussion of grippers, we will focus on the fingers and the mechanical parts.

Several factors must be considered in gripper design. For example, the gripper should be as flexible as the human hand; that is, the gripper should have several fingers that are able to grasp the part. The gripper's fingers should have sensors to detect whether the part is present. The fingers should also be able to bend in several areas, much like the joints of human fingers. And, finally, the gripper's fingers should be able to move independently.

Each of these characteristics is a characteristic of the human hand. Ideally, we would like the robot's gripper to also have these characteristics. But the reproduction of a humanlike hand is too costly for normal industrial production. So, compromises must be made in the design of grippers used in industrial operations. For instance, the dual gripper shown in Figure 11–4B has only two fingers, and the fingers can only move in two ways, up or down.

The main design feature that a user should consider when selecting a gripper is how well it fits the task. For example, the gripper must be able to pick up and move a part from one location to another without damaging the part. The user should also be able to modify the gripper to meet new production demands. For instance, the user might want a gripper designed so that it can pick up a family of parts—that is, parts with the same shape but with different sizes.

A person who is designing grippers should consider several key factors. The following list describes some of these concerns:

1. The part the gripper is to grasp should be within the reach of the gripper. The part should not be hidden from the gripper.
2. The fingers of the gripper should be able to accommodate various sizes of parts and should be self-adjusting.
3. Finger pads (perhaps made of rubber) should be used when the gripper will pick up and place delicate parts.
4. The grippers and the tooling fixtures should work together so that the gripper can reach into the chuck.
5. The gripper must be in the correct position to grasp the part. If the part is heavy, the fingers must grasp the part near or at the center of gravity. If the gripper does not grasp at the center point, the part will tilt in the gripper's fingers.
6. The gripper must grasp the part in the area where the part is the largest.

☰ MECHANICAL GRIPPERS

Many of the grippers for industrial robots are used to transfer parts from one location to another or to assemble parts. These grippers are called **mechanical grippers**. In this section, we will examine the types of mechanical grippers used in industry, the gripping force of the fingers of the gripper, and the drive systems for the gripper.

Types

Mechanical grippers are designed to grasp a part either on the inside diameter or on the outside diameter of the part. Since the grippers must make contact with the surface area, two concerns arise. First, enough frictional force must be applied to the part to overcome the gravitational pull of the part. Second, the gripper must have enough contact force with the part so that when the manipulator rotates, the part will remain in the gripper. The following paragraphs describe several types of grippers that incorporate these features.

Figure 11–6 illustrates a typical **inside diameter gripper**. In Figure 11–6A, notice that the gripper's finger pads are mounted on the outside of the fingers. And in Figure 11–6B, notice that the pads extend below the fingers of the gripper. This mounting allows the pads to fit into the inside diameter of the part that it must lift. The pads are pressed against the inside walls of the part. The frictional force developed allows the fingers to hold the part securely when the gripper lifts the part.

So that the inside diameter gripper holds the part in a secure fashion, several types of contact are available. Figure 11–7A illustrates the typical **two-point contact**. In the two-point contact, two

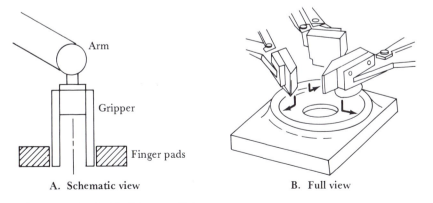

Arm

Gripper

Finger pads

A. Schematic view

B. Full view

Figure 11–6 Inside Diameter Gripper

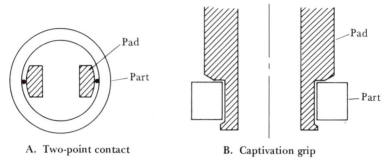

A. Two-point contact   B. Captivation grip

Figure 11–7 Contact for an Inside Diameter Gripper

sections of the gripper make contact with the component. The two-point grip is the best type of contact for the inside diameter gripper.

Figure 11–7B illustrates the **captivation**, or flexible, **grip**. In this type of gripper, the pads of the fingers conform to the inside diameter of the part. These pads are usually made of polyurethane, because polyurethane gives the greatest amount of contact surface between the finger pads and the inside diameter of the part.

Figure 11–8A gives the schematic view of an **outside diameter gripper**. The gripper is designed so that the finger pads press against the outside of the component. The pads develop the frictional force required to lift the part. The pads of the gripper are made from polyurethane bonded to steel, a combination that wears well. Also, the polyurethane can be compressed many times without losing its shape. In addition, polyurethane has a high coefficient of friction, which allows the gripper to grasp various sizes of parts without damaging the parts. Finally, the pads can easily be replaced when they become worn out.

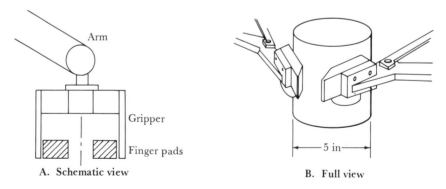

A. Schematic view   B. Full view

Figure 11–8 Outside Diameter Gripper

An outside diameter gripper may be designed to pick up a family of parts, that is, a group of parts with the same shapes but with different sizes. These parts might be pistons with diameters varying from 5 inches to 1 inch. The gripper in Figure 11–8B has the flexibility to pick up both a 5-inch piston and a 1-inch piston.

Sometimes, the design of the part will require that the fingers of the gripper bite into the part. In this case, the fingers of the gripper should be made with hard surfaces and with knurls in the fingertips. For applications involving metal-to-metal contact, hardened steel should be used for the fingers of the grippers.

So that the outside diameter gripper has a positive grip, several types of contact are available. Figure 11–9 illustrates one type, the **four-point-contact vee block**. The vee block gives four points of contact with the surface of the part. Again, the finger pads are made of polyurethane. This type of design ensures that a high coefficient of friction is developed between the fingertips and the part's outside diameter.

A final consideration for mechanical grippers is **compliance**. That is, the gripper should be able to pick up a part even if the part is off-center. Figure 11–10 illustrates gripper compliance. No-

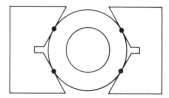

Figure 11–9   Four-Point Contact with Vee Blocks

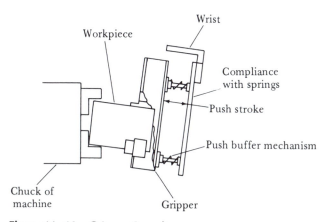

Figure 11–10   Gripper Compliance

tice that springs are mounted between the gripper and a hard stop. When the gripper makes contact with the part, the springs give the gripper some play, or flexibility. That is, if the part is not in the exact programmed position or if the fixture has been moved, then the gripper can locate the part and pick it up.

Gripping Force

The **gripping force**, or the amount of lifting power that a gripper develops, depends on the amount of contact surface between the gripper's fingers and the part. The gripping force should be sufficient to overcome the effects of gravity—that is, the weight—for the part that the gripper is to hold.

The parts handled by a gripper can have various weights. Two factors must be considered in the determination of the weight a gripper will lift:

1. When a part is lifted, the weight of the part will be three times its normal weight. The additional weight is due to the gravitational pull, g, of the earth: $1g$ is due to the weight of the part, and $2g$ is due to the acceleration of the part.

2. When a part is moved in a horizontal plane, the weight of the part is twice its normal weight, because of gravity and acceleration.

Thus, determination of the weight of the part must take into account the gravitational pull of the earth.

The chart in Figure 11–11 can be used to determine the amount of weight that a two-fingered gripper is able to handle when gripping a cylindrical workpiece. For use of Figure 11–11, the dimensions that must be known are as follows (see the inset in Figure 11–11):

— The outside diameter, D, of the part;
— The size, h, of the gripping surface;
— The distance, ℓ_G, from the center of gravity of the part to the center of gravity of the fingers;
— The center of gravity, G, of the part;
— The weight, W, of the part.

The following example illustrates the use of the chart.

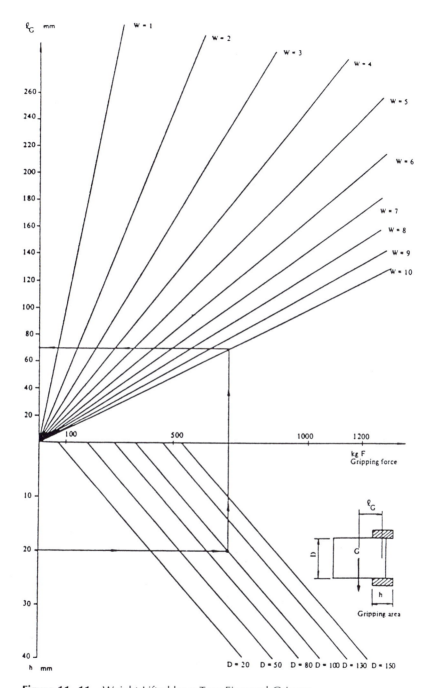

Figure 11–11 Weight Lifted by a Two-Fingered Gripper

Example
A certain part is to be lifted by a two-fingered gripper. The part has an outside diameter of 100 millimeters and a weight of 10 kilograms. The gripping surface of the fingers is 20 millimeters. Determine the distance between the center of gravity of the part and the center of gravity of the fingers.

Solution
Use the chart in Figure 11–11. Draw a box, starting with the known factors. First, use the gripping surface of the fingers, and place a dot at this location. Next, draw a straight line parallel to the horizontal axis. This line should stop at the D = 100 line, as shown in Figure 11–11. Now, draw a vertical line from the D = 100 point to the diagonal W = 10 weight line. From this point, draw a horizontal line to the vertical axis. Read the value on the vertical axis from the chart. The distance between the centers of gravity is 70 millimeters.

Figure 11–12 is a chart that can be used for the same calculation when the robot grips a cylindrical workpiece with three fingers. The methods for reading this chart are the same as those for reading the chart in Figure 11–11. The chart gives the grasping action of the part either on the outside diameter (OD) or on the inside diameter (ID) of the part. The upper part of the chart relates to a vertical grip on the part; the lower half of the chart relates to a horizontal grip on the part.

The box drawn near the axes of the graph in Figure 11–12 illustrates examples of the use of the chart. Consider a vertical grip (top half of the chart). For a gripping force F of 50 kilograms and a gripping area of 25 millimeters or less, the maximum weight W is 10 kilograms. This value (10 kilograms) is read at the point where the top of the box intersects the vertical axis of the chart. Now, consider a horizontal grip (bottom half of the chart). When the workpiece diameter D is 120 millimeters, the distance ℓ_G between the centers of gravity is 26 millimeters or less. This value (26 millimeters) is read at the point where the bottom of the box intersects the vertical axis.

Figure 11–13 is a chart that can be used when the robot lifts rectangular parts. For this chart, the gripper is a two-fingered gripper. Two dimensions must be known for the part that is to be lifted: the length, B, of the part and the center of gravity, G, of the part. The chart also indicates the strength limit rating, P_s, per unit area of the part.

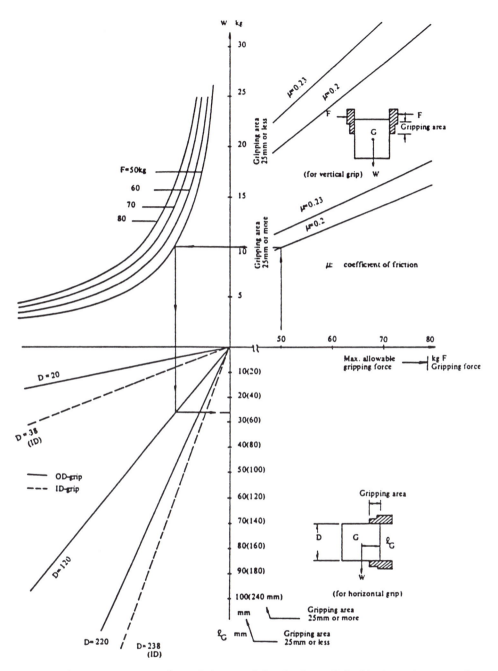

Figure 11–12 Weight and Center of Gravity for a Cylindrical Workpiece and a Three-Fingered Gripper

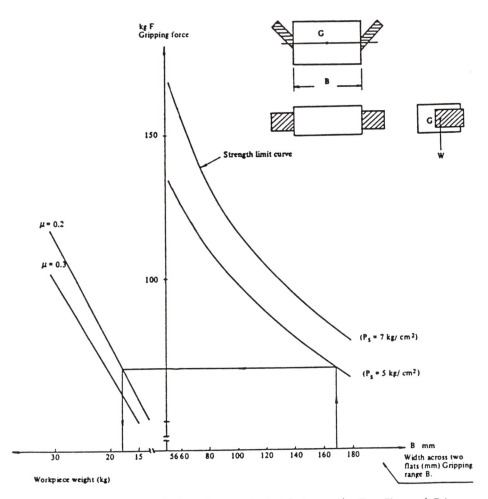

Figure 11–13 Weight for a Rectangular Workpiece and a Two-Fingered Gripper

The box drawn at the bottom of the chart in Figure 11–13 illustrates an example of the amount of weight that this gripper can lift. When the width across the flat of the part is 170 millimeters, the strength of the part is rated at 5 kilograms per square centimeter, the gripping force is 50 kilograms, and the coefficient of friction, μ, is 0.2, the amount of weight that this gripper can lift is about 17 kilograms. This value (17 kilograms) is read at the point where the left side of the box intersects the horizontal axis.

Figure 11–14 is a chart that can be used when the robot lifts a rectangular part with a three-fingered gripper. The dimensions in this chart are the same as the dimensions in Figure 11–12, with

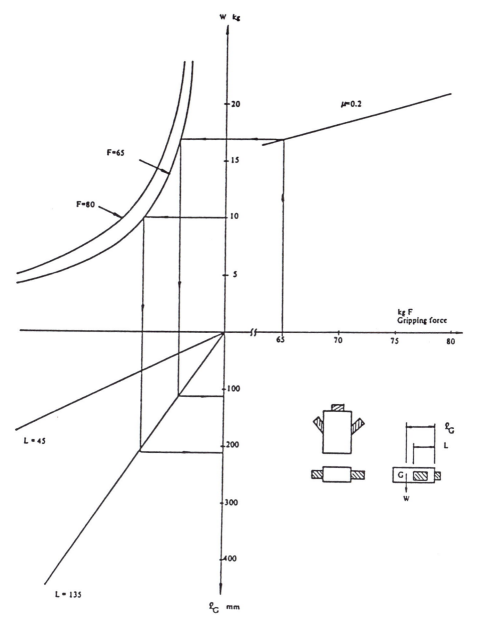

Figure 11–14 Weight and Center of Gravity for a Rectangular Workpiece and a Three-Fingered Gripper

one exception, the dimension L. Dimension L is the distance (length) that the end of the finger should be from the end of the object being lifted.

The boxes drawn near the axes of the graph in Figure 11–14 illustrate examples of the use of this chart. For instance, suppose a gripper has a gripping force F of 65 kilograms and a coefficient of friction μ of 0.2. The length L from the end of the part to the tip of the finger is 135 millimeters. The weight W of the part is 17 kilograms. Then, the distance between the center of gravity of the fingers and the center of gravity of the part is 110 millimeters.

Drive Systems

Mechanical grippers are generally driven by pneumatic systems or gearing systems. Figure 11–15 illustrates a simple gear drive for a gripper. In the figure, the gripper has two axes: the pitch axis, which allows the gripper to move upward and downward around the end effector's flange, and the roll axis, which allows the gripper to rotate clockwise and counterclockwise around the centerline of the gripper. The output gears located at the back of the gripper provide the gripping action for the fingers and provide the movement of the gripper. The linkage bars connect the tips of the gripper to the drive mechanism of the unit. In operation, the motor that drives the gears is activated upon command from the robot's controller. The gears close until the fingers have grasped the part.

Figure 11–16 illustrates a pneumatic drive system for a gripper. The air cylinder develops the action that closes or opens the fingers of the gripper. For this system to operate, a constant pressure must be applied to the pneumatic cylinder.

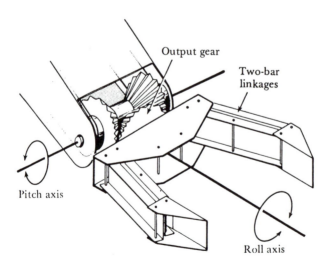

Figure 11–15   Gear Drive for a Mechanical Gripper

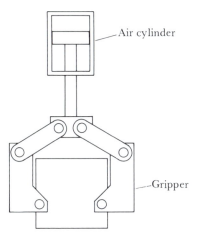

Figure 11–16   Pneumatic Cylinder for a Gripper

VACUUM GRIPPERS

Another type of gripper found in many robotic applications is the vacuum gripper. The **vacuum gripper** uses a vacuum instead of fingers to lift a part. Figure 11–17 illustrates two types of vacuum grippers. Figure 11–17A shows a dual vacuum gripper, and Figure 11–17B illustrates a single vacuum cup for lifting a part.

The vacuum gripper has two components: the cups and the vacuum system. The **vacuum cups** consist of a flexible-rubber cup and a hard-rubber cup. The cup creates negative pressure, which, in turn, creates the vacuum and the necessary lifting power.

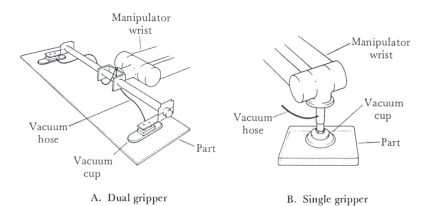

A. Dual gripper   B. Single gripper

Figure 11–17   Vacuum Grippers

The vacuum system is a pump that generates the suction once the cup is in place on the part. Two types of vacuum systems can be used to generate the vacuum: a vacuum pump or a venturi system. The vacuum of a **vacuum pump** is generated by a piston driven by an electric motor. The vacuum pump provides high vacuum pressure at a low cost. The **venturi system** provides high reliability and low cost for start-up. The vacuum system used depends upon the application.

The vacuum cup operates on the principle that a vacuum is created between the cup and the part. This bond causes friction. The friction allows the cup to lift a part. In most applications of lifting, a spring that provides compliance is mounted on the back of the vacuum cup. This spring allows the vacuum cup to come into contact with the part before the cup has reached the programmed position, which ensures that the vacuum cup will develop the proper fit to the part.

The holding force of the vacuum cup depends on the difference of pressure between the outside area of the cup and the inside area of the cup multiplied by the effective area of the cup. We say "effective area" because when the vacuum cup is mated with a part, the cup becomes distorted. This distortion changes the area of the cup.

Vacuum cups are standard, off-the-shelf items that can be purchased by the gripper designer. The number of cups that should be used for an application depends on the weight of the part to be lifted, the size of the cup available to do the job, and the location of the center of gravity of the part. For the best possible results from a vacuum system, the designer should use a large value of pressure difference rather than a large vacuum cup. An important concern in vacuum systems is that a good seal be obtained between the part and the cup.

Once the manipulator has moved the part, it must deposit the part. So, the vacuum system must now be able to release the vacuum. This release is controlled by the robot's program. On command from the robot controller, the vacuum is released, and the part is deposited in the programmed location.

MAGNETIC GRIPPERS

The **magnetic gripper** employs the effect of a magnetic field coming into contact with a ferrous metal. A dual magnetic gripper is illustrated in Figure 11–18. The gripper is made of an electromagnet. A direct, constant current flows through the electromagnet, developing a magnetic field. When the magnetic field of the gripper comes

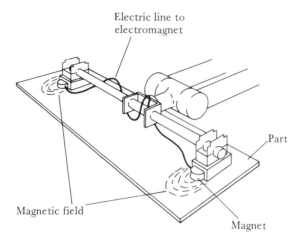

Figure 11–18 Dual Magnetic Gripper

into contact with the part, which is made of ferrous metal, it induces a magnetic field of opposite polarity into the part. This field causes the molecules of the ferrous metal to align and develop smaller magnetic fields in the part. These smaller magnetic fields have a polarity that is opposite to the polarity of the electromagnetic field. Therefore, the two poles of opposite polarity attract each other, allowing the gripper to lift the part.

The electromagnetic current must be turned on and off by an outside DC source. This switching is the job of the program control of the controller.

The magnetic gripper has several special characteristics. First, the magnet's lifting capability must be large enough so that the heaviest part manipulated by the robot can be lifted. Second, the temperature that the electromagnetic gripper can handle is limited to about 140°F. Anything above this temperature will cause the gripper to become ineffective. Special grippers that are able to withstand temperatures of 300°F or more can be designed but at relatively high costs. Finally, the gripper should always make parallel contact with the part. If the part is set at an angle from the magnetic gripper, then a good holding bond might not develop between the magnet and the part. If the magnet is mounted parallel to the part, then the reduction of the magnetic power is only 25%.

The parts to be lifted by the magnetic gripper also have special concerns. For example, parts that have flat surfaces, such as flat, round washers, are the ideal type. Any part that has a large amount of mass will require a large magnet. Also, the surface of the component to be lifted should be smooth and free from any dirt or grease that might be found in the industrial environment. If foreign materials are left on the part, then the bond between the magnetic field and the component will be weakened.

A releasing procedure must also be designed for the magnetic gripper. The release of the part is accomplished by simply reversing the magnetic field through the electromagnet. The magnetic field is reversed by reversing the current flow through the magnet. A switching circuit called a **controlled drop circuit** reverses the current.

The controlled drop circuitry provides another important operation in the magnetic gripper. At the moment that the control drop circuit reverses the magnetic field, it also removes any residual magnetism that has been developed in the ferrous metal part by the magnet. Removal of the residual magnetism is very important in applications of the magnetic gripper. When the electromagnet grips the part for any length of time, the part also becomes magnetized. This magnetism in the part must be removed before the part can be used.

In most cases, the magnetic gripper is able to lift only single sheets of metal that have a thickness of 0.031 inch. So, the electromagnetic gripper may be used with a **separator**, a device that ensures that the gripper will pick up only one part. A separator is illustrated in Figure 11–19. The separator has only enough clearance for one piece of sheet metal to pass through. If additional pieces of metal have been caught by the magnet, they will be stripped away at the separator section.

The magnetic gripper should also have compliance built into it. This compliance may be nothing more than a spring built into

Figure 11–19 Separator for a Magnetic Gripper (Courtesy of GMFanuc Inc.)

the mounting flange between the magnet and the gripper. The spring allows the magnet to settle on the part and ensures that the magnet develops a good holding force for picking up the part.

END-OF-ARM TOOLING

End-of-arm tooling (EOAT) connected to a manipulator performs work on a part. The work performed may be a welding application, a sealing application, a gluing application, and so on. The end-of-arm tooling is mounted to the end effector flange. It requires a safety joint between the tool and the robot manipulator. Again, the safety joint protects the tool in case the robot crashes.

Figure 11–20 illustrates two typical end-of-arm tooling connections. Figure 11–20A shows a drill mounted on the end of the manipulator. The robot positions the tool over the part to be drilled,

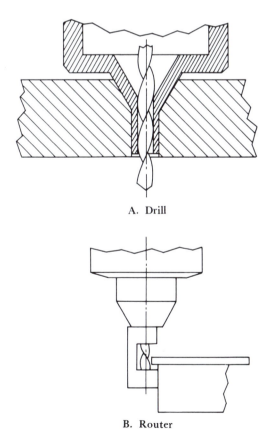

A. Drill

B. Router

Figure 11–20 End-of-Arm Tooling

and a command from the controller's program starts the drilling operation. A template guides the drill to the proper location for drilling. Figure 11–20B illustrates a robot manipulator connected to a router. Again, the program guides the router around the part.

One of the most common jobs for end-of-arm tooling is welding. Figure 11–21 illustrates two typical welding connections. Figure 11–21A shows an arc-welding torch connected to the end effector. The torch is mounted at an angle so that it will be able to reach all programmed points. Arc-welding processes are used mainly for the fabrication and the repair of metal assemblies and for the shaping of metal parts. Figure 11–21B shows a spot-welding gun connected to the end effector. Large industrial spot welders are used extensively in automobile, appliance, and sheet metal fabrication.

The many applications that use end-of-arm tooling cannot be illustrated here. But in each case, the tools are mounted to the robot's end effector. The paths followed by the tool are developed by the robot's path control programming.

SUMMARY

The main function of a manipulator is to perform some type of work. In order for this work to be accomplished, the manipulator must have a device that can perform this work. The devices that perform the work are end effectors. The end effector can be a gripper or an end-of-arm tooling.

Grippers are used to lift or grasp parts. The grippers hold the part until the manipulator puts the part into its programmed position. End-of-arm tooling on a manipulator may be a drill, a router, or some other type of tool. The manipulator guides the end-of-arm tooling through the various steps needed to complete the job.

The end effector is mounted with bolts to the end effector flange on the manipulator. The maximum weight that a manipulator can

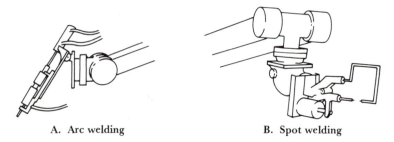

A. Arc welding   B. Spot welding

Figure 11–21 End-of-Arm Tooling for Welding

lift is always measured from the center of the end effector flange and does not include the end effector. Thus, the maximum weight that the manipulator can lift is the weight of the part minus the weight of the end effector.

The operation of the end effector is controlled through a program placed in the controller by the programmer. The interfacing between the end effector and the controller tells the end effector when to perform its task. For example, program control causes a gripper to open and close. For end-of-arm tooling, program control causes a welding gun to turn on and off or a drill motor to start and stop. The addition of an end effector to a manipulator increases the size of the work envelope of the robot. Thus, the length of the end effector must be added to the maximum reach of the manipulator.

The manipulator used in production must be able to meet the various cycle times for performing its operations. The amount of cycle time can be reduced by the use of special dual grippers that are designed to perform two jobs during the same cycle time. Dual grippers allow the manipulator to pick up one finished part and one unfinished part. Effective part presentation also helps to reduce the cycle time of the operation.

In many manipulators, a safety joint is placed between the end-of-arm tooling or gripper and the manipulator. The safety joint, rather than the end effector, is broken if the end effector crashes into a fixture. When the safety joint is broken, an alarm sounds, and the manipulator ceases its motion.

Gripper designs must meet the demands of the various tasks that the grippers must perform, such as holding different-shaped parts. For instance, the gripper should be designed to hold the part without causing damage to the part. Also, the gripper must have fingers that will hold the part in position as the part is moved from one location to another. Finally, the gripper should be flexible enough to grasp a family of different parts.

Grippers can be designed to grasp a part on the inside diameter of the part or on the outside diameter of the part. The gripper should be able to hold the part securely as the manipulator moves into position. The gripper must make contact with the part in several different places. The gripper must also supply enough gripping force to the part to overcome the effects of gravity on the part.

In the design of the gripper, additional weight must be accounted for because of the gravitational pull of the earth and the acceleration of the manipulator. These two factors may mean that the weight of the part is three times its normal weight. Other variables that must be taken into account by the designer are the center of the part to be gripped, how far the gripping point is from the center of gravity of the part, the size of the pads that will grip the part, and the weight of the part to be lifted by the gripper.

Mechanical grippers, which are the type used most often in industrial applications, are driven by either pneumatic actuators or gearing systems. The mechanical gripper is used to lift various sizes and weights of parts.

In many applications, a mechanical gripper will not provide the proper grasping method for the part. In these cases, a vacuum gripper can be used to hold the part in position while the part is transferred from one location to another. The vacuum gripper has rubber cups that hold the part and a vacuum, or negative-pressure, system. The suction created by the vacuum holds the part in position while the manipulator is in motion.

The magnetic gripper is used to pick up parts made of ferrous metals. When the magnetic field of the gripper comes into contact with the ferrous metal part, it induces a magnetic field of opposite polarity into the part. Thus, the part and the gripper attract each other, allowing the part to be lifted.

End-of-arm tooling is a tool connected to the end effector flange of the manipulator. These tools allow the manipulator to perform tasks such as arc welding, spot welding, drilling, routing, deburring, and sealing.

KEY TERMS

captivation grip
compliance
controlled drop circuit
dual gripper
end effector
end-of-arm tooling
fingers
four-point-contact vee block
gripper
gripping force
inside diameter gripper

magnetic gripper
mechanical gripper
outside diameter gripper
payload
safety joint
separator
two-point contact
vacuum cups
vacuum gripper
vacuum pump
venturi system

QUESTIONS

1. For each of the following activities, state whether a gripper or an end-of-arm tool is appropriate: (a) sanding; (b) welding; (c) press loading; (d) sealing; (e) drilling.

2. Where is the end effector connected to the manipulator?

3. A cylindrical coordinate robot has the capacity to lift 25

pounds. A 10-pound gripper is added. What is the maximum weight the manipulator can lift with the gripper?

4. What causes the gripper to return to the same position over and over?

5. A cylindrical coordinate manipulator has a maximum reach of 5.5 feet without the end effector. If a 1-foot, end-of-arm tool is placed on the manipulator, what is the safe work envelope of the manipulator with the end effector?

6. How can cycle time be increased by using a dual gripper as opposed to a single gripper?

7. What process will reduce cycle time with a press-loading operation using a single gripper?

8. What device is placed between the end effector and the manipulator for safety purposes?

9. Which part of the gripper is used to grasp parts?

10. List five concerns that a designer should be aware of when designing a gripper.

11. What type of gripper can be used to pick up a cylinder?

12. What type of material is usually used for finger pads on grippers?

13. If the gripper's fingers are to bite into the surface of a metal, what is the best type of steel to use for the fingers?

14. What factors add weight to the part when the gripper lifts the part from a tabletop?

15. A gripper is to lift a cylinder with two fingers. The part has a diameter of 20 millimeters, the gripping area is 10 millimeters, and the distance from the center of the finger pad to center of gravity of the part is 40 millimeters. What is the maximum weight that this gripper can lift?

16. Which axis on a mechanical gripper allows clockwise and counterclockwise rotation?

17. Which part of the vacuum gripper interfaces with the part?

18. In the vacuum system, what holds the part in position?

19. What factors should be considered for vacuum grippers?

20. Describe the principle behind the magnetic gripper.

21. At what angle should the magnetic gripper come into contact with the part?

22. What is the temperature range for a magnetic gripper?

23. What circuitry reduces the magnetic field in the part that is picked up by a magnetic gripper?

24. What is the most common task assigned to the end-of-arm tooling?

12

Robotic Sensors

OBJECTIVES

Upon completing this chapter, you should be familiar with:

— Types of sensors,
— Proximity sensors,
— Temperature sensors,
— Touch sensors,
— Force sensors,
— Advanced tactile sensors,
— Sensor programming,
— Vision sensors.

INTRODUCTION

For a robot to perform tasks that are now done by humans, the robot must have a sensing ability. The robot employs the **sensor** as a measuring device; that is, it computes the sensor information and acts upon that information. The basic sensors used in the robotics area are devices called transducers. The **transducer** converts a mechanical force to an electric signal. This electric signal is processed by the robot controller, and the robot then performs a job.

Some sensors found on factory floors today are basically limit switches. These switches are either open or closed, which means that the decision capability of the robot controller is very limited. Other sensors allow the robot to perform various assembly tasks and give the robot more decision capabilities. These sensors include force sensors, vision sensors, and tactile sensors. Each sensor expands the robot's usefulness in the workplace.

In this chapter, various sensing devices will be explored as they relate to robotic applications.

TYPES OF SENSORS

Sensors used in robotic applications can be classified as either a contact sensor or a noncontact sensor. Sensors may also be classified as either an internal sensor or an external sensor, and as either a passive sensor or an active sensor. The sensors employed in robotics are primarily contact and noncontact sensors. So, in this section, we will look at these two types.

Contact Sensors

The **contact sensor** allows the robot to sense whether an object is present. This simple type of sensor might be a limit switch placed on a conveyor line, as shown in Figure 12–1A. When the part is presented in front of the robot, the limit switch is contacted, and the robot executes its program. If the limit switch is not contacted, then the robot executes a subroutine in the program for an alarm condition, which alerts an operator for action to be taken.

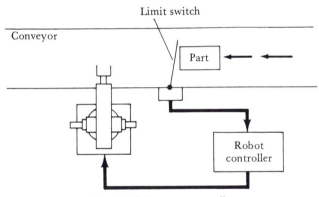

A. Limit switch on conveyor line

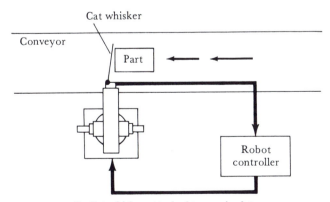

B. Cat whisker attached to manipulator

Figure 12–1   Contact Sensors

Another type of contact sensor is a touch sensor, such as a *cat whisker,* on the end of the robot's arm. This type of sensor is shown in Figure 12–1B. As the robot moves into position, the whisker senses the presence of the part. As soon as the whisker makes contact, the signal is processed by the robot's controller, and a decision is made from the sensor's input to the controller.

The contact sensor measures the response of some contact force. These sensors include touch, force, pressure, temperature, and tactile sensors.

Noncontact Sensors

The second type of sensor used in robotics is the noncontact sensor. **Noncontact sensors** measure the response of components through some type of electromagnetic radiation. This electromagnetic radiation might be the transmission of a light or the development of an electromagnetic field. Noncontact sensors develop a change in the electromagnetic field that is generated by the sensors. This change in strength of the electromagnetic field is then processed by the controller. From this information, the controller directs the robot to perform a task. Noncontact sensors measure temperature changes, electromagnetic changes, and pressure changes.

Figure 12–2 gives an example of a noncontact sensor. An optical device, a light-emitting diode (LED), is attached to the end of the robot's arm. As the robot moves about the part it is to pick up, the sensor detects whether the part is present. Detection is accomplished through the change in the intensity of the LED as it approaches the part. If no part is present, the LED will not change, and the robot will branch to another routine. If the part is present, the sensor indicates this condition, and the robot will continue on its path.

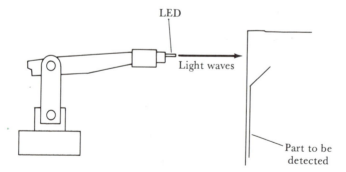

Figure 12–2   LED as a Noncontact Sensor

Another example of a noncontact sensor is a vision system, as shown in Figure 12–3. The vision system allows the robot to locate and pick up different parts. In the figure, the video camera is connected to the end of the manipulator. As the part travels down the conveyor, the camera detects whether a part is present. When a part is detected by the vision system, the vision computer requests the robot controller to branch to a subroutine. The program is called up from the robot controller, and the manipulator is commanded to perform the task described by the program.

PROXIMITY SENSORS

A **proximity sensor** is used to detect the presence of a part when the part is within the range of the sensor. The proximity sensor can be either a contact sensor or a noncontact sensor. The contact proximity sensor determines whether a part is present when the part comes into contact with the sensor. The contact sensor is generally a limit switch that closes when it contacts the part. The noncontact proximity sensor does not have to come into contact with the part in order to determine its presence. With a noncontact sensor, the detection of a part is generally accomplished through the use of a magnetic field. Both types of proximity sensors are discussed in the following subsections.

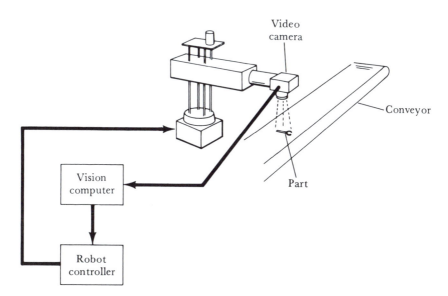

Figure 12–3 Vision System as a Noncontact Sensor

Limit Switches

Figure 12–4A illustrates a limit switch used as a sensor. The **limit switch** is a normally open, single-pole, single-throw switch. This switch has large current ratings and voltage ratings for industrial applications. The switch is built into the end of the robot's arm. As the robot approaches the part, the limit switch closes. The closed contact transfers a signal to the controller. The controller receives the signal, and from the on or off condition of the signal, the controller branches to another part of the program. In Figure 12–4B, the program flowchart illustrates the contents of this program. Notice that if the sensor does not transmit a signal to the controller, the robot will execute another part of the program (move to perch position) while waiting for the contact switch to be closed.

Electromagnetic Sensors

Electromagnetic proximity sensors are generally light source proximity sensors. Through the use of light-emitting diodes, these sensors detect the distance the robot's arm is from a part.

Figure 12–5 illustrates an **LED sensor**. Notice that the sensor has two components: a transmitter and a receiver. The LED is the light-transmitting source, and the phototransistor is the receiving source. The operation of these two components together forms the proximity sensor.

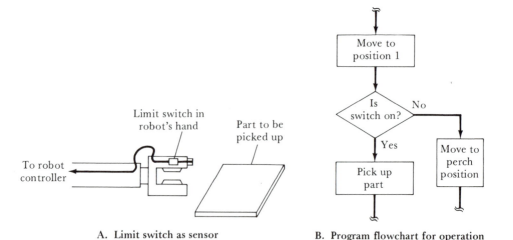

A. Limit switch as sensor   B. Program flowchart for operation

Figure 12–4 Limit Switch and Program Flowchart

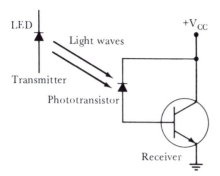

Figure 12–5   LED and Phototransistor as a Proximity Sensor

In operation, light from the LED falls on the input of the phototransistor, and the amount of conduction through the transistor changes. The less light that falls on the transistor, the less it will conduct. And the more light that falls on the transistor, the more it will conduct. The controlling factor is the light generated from the LED and the receiver circuitry of the phototransistor.

In application, this proximity sensor is used as follows: As the arm of the robot reaches down to pick up a part, the intensity of the light received by the phototransistor changes. If a part is present, the amount of light received by the phototransistor increases. This increase in light is converted into a signal and then processed by the controller, and the robot executes a command. If the part is not present, the light received by the phototransistor decreases. Thus, the phototransistor does not conduct. This signal is then processed by the controller, which performs a routine of no part present from this information.

One of the major drawbacks of LED proximity sensors is that dirt or foreign objects in the transmission or receiver path can block the light. Hence, the sensing function will be lost. So, the programmer might want to include an alarm condition in the program that will sound a siren if the sensing function fails. In the alarm condition, the robot will stop operation and automatically move to a safe location.

Another type of electromagnetic proximity sensor is the **Hall effect sensor**. Figure 12–6 illustrates the basic operation of the Hall effect proximity sensor. As illustrated in the figure, the Hall effect detector operates on the effect of a current passing through a semiconductor material.

In Figure 12–6A, the current passing through the conductor is constant, because no external magnetic force is applied to the semiconductor material. Thus, the output voltage is 0 volts. The 0-

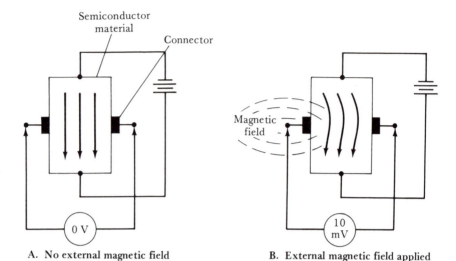

A. No external magnetic field   B. External magnetic field applied

Figure 12–6   Hall Effect Sensor

volt output from the sensor tells the controller that no part is present. In this case, the controller may have several options. For example, the controller may have the robot wait at the location until the part is present, or the controller may move the manipulator to a perch position and generate an alarm.

In Figure 12–6B, a magnetic field, generated by the current passing through the semiconductor material, is brought close to the material. As the magnetic field approaches the material, the current in the semiconductor changes. This change develops a voltage. The voltage generated as the output of the Hall effect sensor is typically only about 10 millivolts. This small amount of voltage is then processed by the controller, and a program routine is performed.

The Hall effect sensor's main task is to sense a change in the magnetic field.

TEMPERATURE SENSORS

In many applications, a robot is used in high-temperature areas that are hazardous to humans. Such a robot has a **temperature sensor**, or probe, attached to the end of its arm. The temperature probe measures the temperature of the environment around the robot. Typically, the temperature-sensing device is a thermocouple or a resistance temperature detector. These devices are discussed in the following subsections.

Thermocouples

The **thermocouple** is the basic temperature-sensing device. As shown in Figure 12–7, the thermocouple is made from two types of metals joined together. As the temperature around these metals changes, the thermocouple generates an analog voltage. That is, as the temperature of the metals' junction increases, the voltage increases. The range of voltages outputted from these devices depends on the type of thermocouple used. For example, for an upper limit of heat of 4500°F (2500°C), the output voltage from the thermocouple can be as large as 5.4 millivolts. The small signal voltage generated from the thermocouple has to be processed through a signal amplifier to increase the signal strength.

A temperature probe may be connected to the end of a robot's arm. In this case, the robot has two functions. One function is to place a part into the furnace for heat treating. The other function is to monitor the temperature of the furnace. If the temperature rises above the limit, the sensor communicates that information to the controller. The controller then processes a subroutine that removes the robot's arm from the furnace and moves the robot to a home position. Also, the controller generates an alarm condition and a call for service from the operator.

Resistance Temperature Detectors

Resistance temperature detectors (RTDs) are transducers that can be used to sense heat changes. The RTDs are manufactured from pure metals such as platinum, nickel, tungsten, and copper. Each of these metals develops a positive temperature coefficient. Thus, the relationship between the temperature and the metal's resistance is a proportional relationship. As the temperature increases, the resistance of the metals increases, as shown in Figure 12–8A.

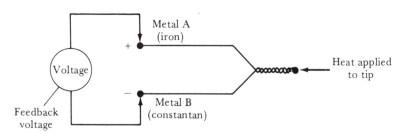

Figure 12–7   Thermocouple Sensor

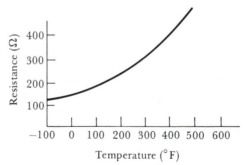

A. Temperature versus resistance change

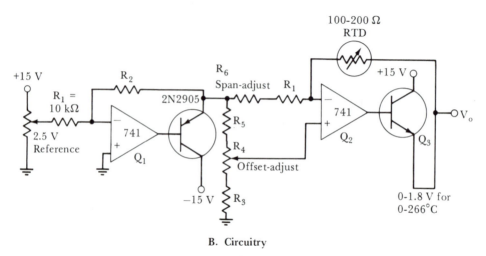

B. Circuitry

Figure 12–8 Resistance Temperature Detector

The RTD can be wired into a bridge circuit that provides accurate measurement of temperature changes. Figure 12–8B illustrates a typical circuit that can be used for detection of temperature
change. The change in resistance developed by the RTD is very
small, and this change must be amplified in order to be detected.
In the figure, two operational amplifiers are used to amplify the
small signal changes that develop in the circuitry. The RTD output
leads are connected as part of the feedback circuitry for the Q_2
operational amplifier. Any change that develops in the RTD changes
the amplification of Q_2.

Operational amplifier Q_1 is used to establish a reference voltage in the circuitry. This reference voltage is established through
the 2.5-volt reference adjust and resistor R_1. Resistors R_2 and R_1
establish the gain for Q_1, which maintains a constant voltage to
the base of the 2N2905 transistor. This constant voltage then es-

tablishes a constant voltage for the voltage divider network of R_3, R_4, and R_5. This reference voltage is then fed to the positive terminal of Q_2.

Any change that is detected by the RTD is fed back to the negative terminal of Q_2 and compared against the reference voltage established on the positive terminal. A difference between these two voltages is amplified and sent to the output transistor Q_3. These small changes in voltage can be related to changes in temperature. The typical range of output voltage is from 0 to 1.8 volts for a change in temperature of 0° to 266°C.

≡ TOUCH SENSORS

The **touch sensor** is used to detect the robot's contact with an object in its path. The touch sensor is basically a limit switch that is mounted on the end of the robot's arm. The limit switch tells the robot's controller that the robot has come into contact with an object by sending back a closed-switch signal.

An application of the touch sensor is shown in Figure 12–9. In this application, the robot is inspecting different parts. The limit switch is attached to the end-of-arm tooling, and the switch is connected to a rolling wheel. As the wheel rotates over the part, the wheel detects any faults on the linear path of the part. If the part has not been machined properly, dents or bumps are detected by the wheel. The limit switch then closes, sending a message to the

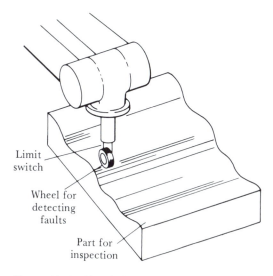

Limit switch

Wheel for detecting faults

Part for inspection

Figure 12–9   Touch Sensor

controller. The controller is then directed to an error subroutine of the program. This subroutine commands the robot to pick up the part and deposit it in the defect basket.

The touch sensor may also be used to keep the robot's arms from touching each other, to ensure that the robot's gripper is touching a part, or to help the robot reach a target or identify a part.

FORCE SENSORS

Force sensors can be used to measure the amount of downward force the robot arm is applying to a part or the amount of force that the robot is using to screw in a bolt. Force sensors for a robotic system are generally of two types: the strain wire gage or the semiconductor strain gage. Both types are discussed in the following subsections.

Strain Wire Gage

The **strain wire gage** measures the strain that is placed on an object. The gage converts a mechanical strain to an electric signal. Figure 12–10 illustrates a typical strain wire gage for a robotic system. The force applied to the gage causes the gage to bend. This bending action also distorts the physical size of the gage. This distortion develops a change in the resistance of the metals of the gage. This resistance change is fed to a resistance bridge circuit that detects small changes in the gage's resistance. The output from the bridge is then inputted to an amplifier circuit much like the circuit for the RTD sensor.

In applications, the strain wire gage can be connected to the end of the robot's hand. As the robot applies pressure to a metal, the strain gage monitors the amount of force developed by the robot's hand. If the force applied exceeds the limit, the controller will branch to another routine to stop the robot's motion.

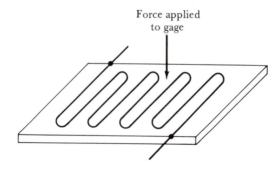

Force applied
to gage

Figure 12–10 Strain Wire Gage

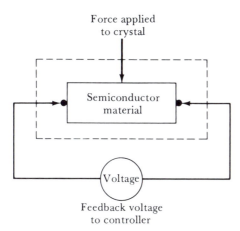

Figure 12–11   Semiconductor Strain Gage

Semiconductor Strain Gage

The **semiconductor strain gage** operates on the same principle as that of the strain wire gage. As a change in the physical shape of the material is developed, the resistance of the material changes.

Figure 12–11 illustrates a semiconductor strain gage. This gage is made from a semiconductor material, a piezoelectric crystal (common rock salt). As a force is applied to the crystal, the shape of the crystal changes. This change of shape develops a voltage at the output, which is then fed to the robot controller. The semiconductor strain gage is used when rapid changes are taking place on the measuring device and when high sensitivity is needed.

ADVANCED TACTILE SENSORS

Advanced tactile sensors give the robot the ability to feel with the gripper. Tactile sensors are generally microswitches placed in the fingers of the gripper. These sensors tell the controller if, for example, a part has been picked up by the gripper or if there are any rough spots on a casting that the robot has deburred.

Figure 12–12 illustrates a gripper that uses tactile sensors. The fingers of the gripper are coated with many microswitches. The **microswitch** is a small, normally open switch. Microswitches are the same switches that are found on a calculator's keypad. If the switches detect a part, this data is fed to the controller. From this data, the controller decides whether a part is held by the gripper or whether the part is offset from the center of the gripper. The controller then has the robot adjust the amount of force used to hold the part.

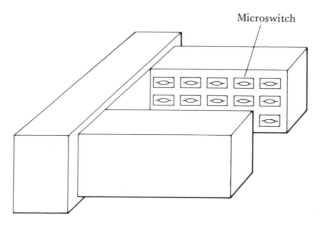

Figure 12–12 Tactile Sensors in a Gripper

Tactile sensors could allow the robot to feel various parts in a bin and select the correct part. This task requires a very complex computer system to store all of the shapes that the robot's hand will encounter. When the robot grasps a part, the sensor feeds the data to the computer, the data are compared with data in memory for identification of the correct part.

If the sensing power of a human fingertip were to be matched by a robot, the fingers of the gripper would have to contain at least 10,000 sensing elements per square centimeter. A 10 × 10 microswitch array placed in the fingers of the gripper would give the fingers the same sensing power as human fingertips. And each of the sensing elements of the gripper would have to be scanned by the computer every 10 microseconds. The force these fingers would sense would have to range from 1 gram to 1000 grams. And if the grippers were used to lift parts, they would have to be made from material that is similar to human skin; that is, they would have to be able to grasp and hold on to parts by the natural friction of the material. These tactile sensors are components of the future. But as technology advances in robotics, tactile sensors will become a common component.

SENSOR PROGRAMMING

As we have seen, the sensor identifies the part's location, and the controller must adapt to this situation by adjusting the axis of the robot's motion. So, in sensor applications, program control is a very important operation. **Program control** is responsible for reading the data that is transmitted by the sensor and then acting on that information. In high-technology controllers that use sensors, there are two important steps that must be performed by the controller's program. These steps are reading the sensor's data and developing positional offset data to compensate for the sensor's data. These two steps are described in this section.

Reading Sensor Data

The controller must read the sensor's data and then act on that data. Figure 12–13 illustrates a simple program flowchart used for reading sensor data. The sensor used in this application is a limit switch that detects whether a part is present.

In step 1 of the program, the manipulator is moved to a perch position away from the part presentation station. The programmer has selected register 10 to keep track of whether the part is present. If the part is present, a 1 is placed in register 10. If the part is not present, a 0 is placed in register 10. The program designer has also used the first step for a housekeeping routine that clears the register every time the manipulator goes through its cycle.

In step 2, the manipulator is moved to position 1, which is directly in front of the manipulator. In step 3, the controller activates the sensor. In step 4, the sensor's data is read by the controller, and in step 5, the contents of the sensor are placed in register 10.

Step 6 is a decision block that determines whether a part is present. Remember that a 1 means that a part is present, and a 0 means that no part is present. If a part is present, the program continues to step 7, where the manipulator picks up the part. In step 8, the manipulator palletizes the part. Step 10 is used to keep track of the number of times this operation has been done. That is, when register 5 equals 10, the operation will stop. If register 5 does not equal 10, then the program branches back to step 1 and repeats the cycle.

Step 9 is a very important step. In this step, if a part is not detected by the sensor, the program jumps to a halt routine. In this routine, the manipulator moves to the perch position away from the part presentation station and generates a user's alarm. This alarm will sound at the operator's workstation. The operator can then go to the robotic work cell to correct the problem.

Reading Positional Offset Data

The sensor also develops positional offset data for the operation of the program. Positional offset data is different from the positional

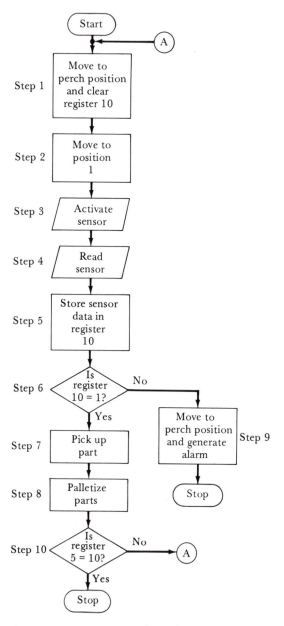

Figure 12–13 Flowchart for a Sensor Used to Detect a Part's Presence

data that has been programmed into the controller. **Positional offset data** tells the controller the difference between the actual position and the programmed position of the part the robot is working with. Positional offset data can be developed by a single point of data or by two or more points.

Figure 12–14A illustrates a positional offset example for a single point of data. In this example, the sensor is to detect the location of a part that the manipulator must weld. Originally, the robot controller was programmed with data for a programmed position of the part. The programmed position is shown by the dotted lines in Figure 12–14A. One point of the location, point A, was programmed into memory. The part has now shifted to a new location, illustrated by the solid line and point A'.

As the manipulator approaches the part, the sensor looks for point A but finds that the position has been shifted to point A'. The sensor measures the distance D the part has shifted and records this data in the controller's operational register. As the manipulator approaches the part, the controller's program reads the offset position stored in the register and compensates the manipulator's position to point A'.

The major drawback with this type of sensing circuitry is that a vision system must be incorporated to detect the change in position. Also, the vision system only detects one point, which means that the part can only be out of position in one direction for this system to function properly.

The limitation of the sensor system shown in Figure 12–14A can be overcome with a sensor that can read three points of the position of the part. Figure 12–14B illustrates this type of offset positional reading. In this example, the position of the part has three reference (programmed) points, points A, B, and C. The part has shifted from the programmed position, illustrated by the dotted lines, to the new position, illustrated by the solid lines.

Again, a vision system is employed for measuring the offset data. As the manipulator approaches the part, the vision system reads the three points A', B', and C' and compares them with the programmed points stored in memory. The differences between the three programmed points and the three offset points are measured in X, Y, and Z offsets. These offsets are then compared, and the movement of the manipulator is adjusted.

VISION SENSORS

Vision is one of the most important human senses. Our eyes allow us to see color and to sense motion. The eye is made up of hundreds of millions of rods and cones in the retina. These rods and cones are nothing more than sensors. From these sensors, the eye receives many different bits of information and, through the brain, is able to convert these small pieces of input into a picture. Since there are a great number of sensors, the eye is able to develop great detail in the picture. The human eye also has the ability to see a part for

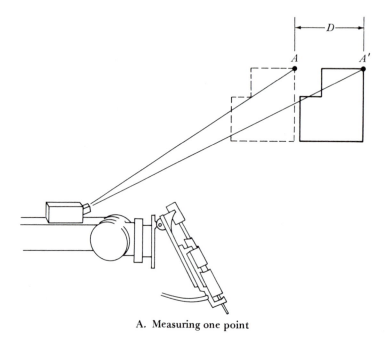

A. Measuring one point

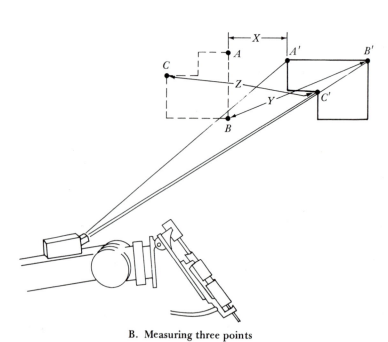

B. Measuring three points

Figure 12–14   Reading Positional Offset Data

the first time without previously being taught about the part. And it can tell the difference between light and dark areas of the object.

One of the major research areas in robotics is in **vision sensors**. The vision systems used today employ a sensor system and a computer-operated memory bank. The sensor part of the system is handled by a device called a videocon tube or a solid-state video camera. The memory bank is handled by a computer system that has enough memory capacity to store all the objects that the robot will have to view.

Video Input Images

To understand the process of programming a picture for a computer, we will consider an illustration called a halftone. This type of illustration is used, for example, in photographs printed in newspapers. The **halftone** is made up of different areas of light and dark and of many different dot sizes. These dots form a matrix, so the picture is also called a **dot matrix picture**. The size of the dot determines the brightness of the picture at that point. The difference between the light and the dark areas of the picture is called a picture cell. Each of the picture cells in the matrix is identified by a level of grey.

The picture information about differences between the light and the dark areas comes from a **video camera**. The video camera converts the light from an image into electric signals. The video camera is a transducer, because the video camera converts one form of energy into another form. In many video cameras developed for the vision system in robotics, a new device called a **charge-coupled device** (CCD) is used to convert the light images from the picture into electric signals. The CCD uses large-scale integration (LSI) technology for the conversion.

The CCD is mounted behind the lens of the camera. The images are transferred through the lens to the CCD unit. The different levels of brightness are then converted into digital pulses by the electronics of the camera and are transferred to the vision computer system. In the vision computer system, the image is stored in memory. The light and dark areas of the images are broken down into bits of information. These bits of information are 1s and 0s. The 1 indicates a high, or bright, level, and the 0 indicates a low, or dark, level.

Each of the different lines of the picture must be broken down by the camera's electronics for storage into memory locations of the computer. Figure 12–15A illustrates a typical scanning process used. The CCD unit is scanned by the electronics of the camera to

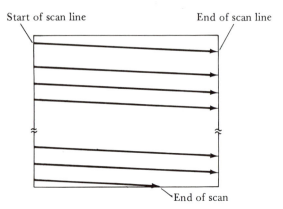

A. Scanning by the electronic beam in the camera

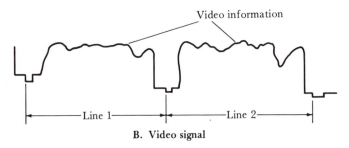

B. Video signal

Figure 12–15   Vision Scanning Process

break the image of the CCD into lines of image information. The scanning of the CCD starts in the upper left-hand corner of the display and goes from the left side to the right side of the unit. These lines of the picture are then converted into digital pulses and stored in the memory of the vision computer. The scanning process continues from the top of the picture to the bottom. Each time a horizontal scan is completed, the camera returns to the left edge and scans the next line. One drawback to these types of cameras is that each line is scanned one by one. If a pixel in the line is not operational, the scanning process will stop. This will stop the camera from scanning the entire picture, and data will not be received from other pixels.

Figure 12–15B illustrates a typical video signal developed from the camera. The signal is made up of different levels of voltage. Each voltage level represents a level of grey or white.

As the camera scan reaches the right edge of the picture, a pulse tells the camera to return to the left side of the next line. This pulse is a **synchronizing pulse**. Its main function is to make the camera scan the picture in the proper sequence. The synchro-

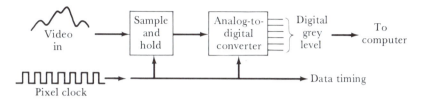

Figure 12–16   Pixel Clock for Converting a Video Signal to Digital Bits

nizing pulse is also used by the video display terminal to keep the display's scanning rate the same as the camera's scanning rate.

The synchronizing pulses for the vision system are developed by a **pixel clock**. The pixel clock has two functions. One function is to time the scanning of the CCD unit in the camera. The second function is to time the conversion of the video signal into a digital signal for the vision memory.

Figure 12–16 illustrates the use of a pixel clock in the digital conversion of the video input signals. The video input signal is fed into a sample-and-hold circuit. The pixel clock data is also fed into the sample-and-hold circuit. The video signal is then transferred to the analog-to-digital converter (ADC) circuit on signal from the pixel clock. The output of the ADC is then sent to the vision's computer memory. The bits of digital information that have been converted are now stored for later referencing by the computer.

Distinguishing an Object

The main function of the vision system in robotics is to locate and distinguish objects or parts. The objects' images have already been identified by the video camera, converted to digital signals, and stored in the computer's memory. This process is training the vision system. The next task is to have the robot, through the vision sensor, distinguish one part from many different parts located on an assembly line. Two methods are used for distinguishing an object: the edge detection process and the clustering process. Both methods are discussed in the following paragraphs.

The **edge detection process** uses the concept of the difference between the light and dark areas of an object. The camera, mounted on the manipulator, locates an object. The image of the object is broken down into digital pulses, which are sent to the vision computer. The vision computer searches through its memory, comparing the different areas of the image with the areas already stored in memory. When the vision computer finds a match between the areas that contain a high difference in contrast, the vision computer has located the edge of the object.

The main problem with the edge detection process is that the vision system is subject to noise spikes. **Noise spikes** are large electric signals that develop in the lines that connect the video camera with the vision computer. The spikes are generated when electrons in the system crash into one another or when the electronics is passed through an integrated circuit. Noise spikes cause the various samplings of the images to be different, and therefore, they give the computer incorrect information. The edge detection process also requires that the vision system operate in an area that is well lit and that the camera be able to detect high levels of contrast between the various elements of the image being viewed.

In the **clustering process**, the video input signal is processed by the camera and fed to the computer's memory. The computer compares the different images in order to match the various contrast levels of the object being viewed. The chain of images that match in contrast levels traces the lines of the object. This process forms a picture of the object. That is, when the levels of contrast match, the object has been distinguished by the vision system.

The clustering process must be done in a well-lit area where the contrast between the light and dark areas can be picked up by the video camera. And, once again, the noise from the surrounding areas will distort the video information.

Template Approach to Image Recognition

Both the clustering method and the edge detection method have difficulty in the comparison routines. Many of the objects that must be identified by the computer will not always have the same orientation. But the vision system must be able to view these objects and recognize them as being the same part.

Figure 12–17 illustrates the **template approach** for part recognition. In Figure 12–17A, notice that the part stored in memory and the part being viewed have the same orientation. The edge detection process can locate the part in this case.

In Figure 12–17B, the part viewed by the camera has been rotated by some angle. In the template approach, the computer rotates the image so that it can be identified and matched. This process involves a great amount of computer memory.

Vision Algorithm

The **vision algorithm** is the computer program employed for the vision system. The vision algorithm is currently being worked on by many research companies and universities around the country.

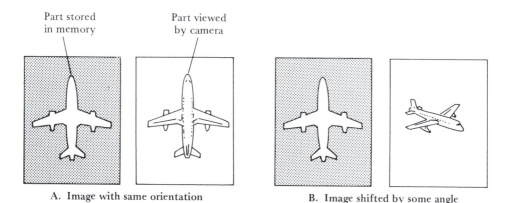

A. Image with same orientation   B. Image shifted by some angle

Figure 12–17   Template Approach for Image Recognition

One such algorithm has been developed by the Stanford Research Institute. The SRI algorithm uses the fact that the images of an object are made up of different shades of light and dark areas. The light and dark areas correspond to 1 and 0 in the digital signal of the vision system. For this algorithm to operate, the camera must see the entire silhouette of the object. The algorithm then compares the stored information in computer memory with the silhouette of the object. If the silhouette and the computer's stored image match, the robot is commanded to pick up the part.

Another vision algorithm and a programming language for it have been developed by the Automatix Company. Their programming language is called the Robot Automatix Incorporated Language (RAIL). Figure 12–18 illustrates a typical program developed for the RAIL system. This program uses a vision system to inspect IC chips as they are processed on an assembly line. The main function of this program is to detect whether the part is within tolerance.

The program for the vision system is made up of an IF–THEN–ELSE statement. The program simply states that if the maximum value of X, the heat sink of the chip, is greater than or equal to the offset values stored in the computer's memory, then the part is OK. This simple algorithm is employed in the inspection process that uses the vision system.

Two important steps in the program are the XMAX and the ORIENT steps. These two commands tell the system to search for the components that are important to the offsets of the program. The XMAX step identifies the location of the heat sink of the IC chip in relationship to the IC chip itself. The ORIENT command identifies the orientation of the part, such as whether the part is

```
INPUT PORT 1 : CONVEYOR
INPUT PORT 2 : PART_DETECTOR
OUTPUT PORT 1 : BAD_PART

WRITE "ENTER CHIP OFFSET TOLERANCE: "
READ OFFSET_TOL
WRITE "ENTER CHIP TILT TOLERANCE: "
READ TILT_TOL

WAIT UNTIL CONVEYOR = ON

WHILE CONVEYOR = ON DO
  BEGIN
    WAIT UNTIL PART_DETECTOR = ON
    PICTURE
    IF XMAX {"HEAT SINK"}-XMAX {"CHIP"} > = OFFSET TOL
          AND
        ORIENT {"CHIP"} WITHIN TILT_TOL OF 90
        THEN
          BAD_PART - OFF
        ELSE
          BAD_PART - ON

  END
```

Figure 12–18 RAIL Program for Vision Systems

shifted by 90° or more. The final steps of the program identify whether the part is good or bad. If the part is good, the robot advances the conveyor line. If the part is bad, the robot removes the defective part from the conveyor.

SUMMARY

Sensors that are used in robotics applications are measuring devices. These devices include force, vision, and tactile sensors. The sensors for robotics applications are transducers. Transducers convert one form of energy into another form. Robotic sensors can be classified as a contact sensor, which must come in contact with the object to sense it, or a noncontact sensor, which measures the response of components through electromagnetic radiation.

A proximity sensor detects the presence of a part. These sensors can be classified as either a contact sensor or a noncontact sensor. A contact proximity sensor is a limit switch. A limit switch mounted on a manipulator or a conveyor identifies that a part is present as soon as the part makes contact with the limit switch. The noncontact proximity sensor is an electromagnetic radiation device. As the object comes into contact with the magnetic field, the strength of the field changes.

Temperature sensors are used when the manipulator must work in areas that have high temperatures. A temperature sensor may be a thermocouple or a resistance temperature detector. The thermocouple is made from two different metals joined together. As the temperature changes around the thermocouple, the current generated by the thermocouple also changes. In a resistance temperature detector, the resistance of the RTD changes as the temperature changes.

A touch sensor is used in many robotics applications to identify whether a part is present. The basic touch sensor is a simple limit switch that is open or closed depending on the presence of a part. A touch sensor can also be used in the inspection area to detect whether a part has any defects.

Force sensors are devices that measure the amount of force that a manipulator applys to an object. The force sensor is basically a strain gage. In the strain gage, the bending of the metal changes its resistance, and this change is used to measure the strain. The strain gage can be a wire gage or a semiconductor gage.

Advanced tactile sensors are sensors that use microswitches in the tips of the robot's finger pads. These switches allow the gripper to detect the presence as well as the shape of the part. These sensors allow the gripper to become very much like the human hand.

The information that a sensor detects must be fed back to the controller so that the controller can act upon the information and correct the operation of the manipulator. The data developed by the sensor is communicated to the controller through program control. Program control also includes positional offset data, which is data that tells the controller the difference between the programmed position and the actual position of the part. Positional offset can be identified through a single point or up to three points of positional information.

The vision system for a robot allows the robot to select various parts presented to it and to provide offset information to the controller and the program. The images are viewed by the vision system through a video camera that contains a charge-coupled device. The CCD converts these images into binary bits of information. These bits are stored in the vision computer memory for later recall in part recognition.

Many of the vision operations require a vision algorithm. The algorithm is the program that converts the information from the camera into useful data for the computer. One of the first vision algorithms developed was the SRI algorithm.

310

KEY TERMS

advanced tactile sensor
charge-coupled device
clustering process
contact sensor
dot matrix picture
edge detection process
electromagnetic proximity
  sensor
force sensor
halftone
Hall effect sensor
limit switch
microswitch
noncontact sensor
pixel clock
positional offset data

program control
proximity sensor
resistance temperature
  detector
semiconductor strain gage
sensor
strain wire gage
synchronizing pulse
temperature sensor
template approach
thermocouple
touch sensor
transducer
video camera
vision algorithm
vision sensor

QUESTIONS

1. Name two classes of sensors found on robotic systems.

2. What type of device is used in the sensor to convert the energy from one form to another?

3. What is one of the drawbacks in using an LED as a proximity sensor?

4. Discuss the operating principle of the Hall effect sensor.

5. What is the typical output voltage developed from a Hall effect sensor?

6. What is the upper limit for the temperature and for the voltage generated from a thermocouple?

7. What type of temperature coefficient is developed by an RTD?

8. What is the typical voltage range generated from an RTD?

9. Name a typical application of the touch sensor.

10. Describe the basic principle of operation of the strain wire gage.

11. What type of material is used for the semiconductor strain gage?

12. What type of component is used in an advanced tactile sensor, and where is it found?

13. What three parts of the robotic program can the sensor adjust?

14. Define the term *positional offset data*.

15. Where is the information that is read by the sensor stored in the controller?

16. Which part of the vision system converts the image into digital information?

17. What is the difference between the light and dark areas of a picture called?

18. Which part of the vision system is the transducer?

19. Which vision process detects the difference between the light and dark areas of the image?

20. What is the main difference between the edge detection process and the template approach to image recognition?

21. Define the term *algorithm*.

13

Robotics Applications

OBJECTIVES

Upon completing this chapter, you should be familiar with:

— Material handling,
— Machine loading and unloading,
— Die casting,
— Welding,
— Inspection,
— Assembly,
— Spray painting.

INTRODUCTION

Many of the applications for the robot are in the industrial area. In the past, robots were used primarily in applications where human risk factors were very high. For example, the first application of the robot was in nuclear power plants. Robotic arms were used to change the uranium rods in the nuclear reactor. Today, robots find many other applications, such as work that is very repetitive. For example, in spot welding, robots are employed to generate welds along a taught path. Robots are also used in material handling, where material from an assembly cell must be stacked in various locations. Through the use of sensors, the robot can detect if the part has completed its cycle and is ready for stacking. This stacking process is called palletizing the parts.

Many routine tasks that in the past have been done by the human worker can now be done by the robot. The robot frees human workers from hazardous situations and from boring tasks.

In this chapter, several robotics applications will be explored. Also, we will see how robots can be integrated into a manufacturing facility.

===== MATERIAL HANDLING

One of the most common jobs that a robot can be adapted to is a material-handling operation. **Material handling** is a process in which the robot transfers parts from one operation to another operation. For instance, the robot might take the parts from a conveyor line and deposit them at some type of press.

Figure 13–1 illustrates an overhead view of a material-handling operation. Parts are supplied to the manipulator by the conveyor line. The manipulator places the parts on the overhead monorail. Notice that safety fences are placed around the work envelope. So, while the manipulator is in operation, operators do not have access to the work envelope. If the work envelope is breached by the operator, the cell automatically shuts down for safety reasons. The work cell also contains an area in which the operator can monitor the transfer operation.

For material-handling operations, the robot must have the following features:

— The manipulator must be able to lift the part from the conveyor to the press.
— The robot must have the reach needed.
— The robot must be a cyclindrical coordinate robot.
— The robot's controller must have a large enough memory to store all of the programmed points so that the robot can move from one location to another.
— The robot must have the speed necessary for meeting the transfer cycle of the operation.

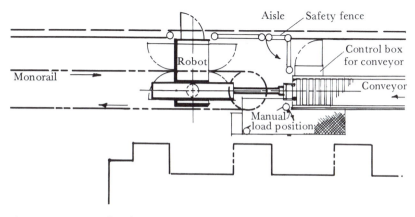

Figure 13–1 Overhead View of Robotic Parts Transfer

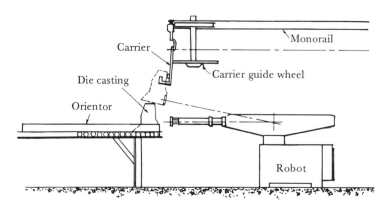

Figure 13–2   Machine Transfer Work Cell

Figure 13–2 illustrates some of the design considerations that must be taken into account when one is developing a machine transfer work cell for the robot. For instance, the robot's upward and downward stroke must be long enough to reach the orientor and the press input from the carrier. The orientor moves the part into the correct position for its travel down the conveyor line. In this application, the robot takes the die casting from the monorail carrier and places it on the orientor. So, the robot must also have the lifting capacity to lift the die casting from the carrier and transfer it to the orientor line. Another important concern is that the robot controller must be in sequence with the monorail speed. The rail speed determines when a die casting is available for the robot.

In the following subsections, we will examine two typical material-handling operations for a robot, namely, palletizing and line tracking. We will also see how the tasks in a material-handling operation fit together for process flow.

Palletizing

Many material-handling applications require the robot to stack parts, that is, to **palletize** them. There are two processes involved: palletizing and depalletizing. Palletizing, for example, could be the process of taking parts from an assembly line and stacking them on a pallet. Depalletizing, for example, could be the process of taking the parts off the pallet and placing them on the assembly line.

Figure 13–3 illustrates a typical palletizing operation that uses a cylindrical coordinate manipulator. The bottles are presented in front of the manipulator at a parts presentation station. The parts presentation station ensures that the bottles are in the proper

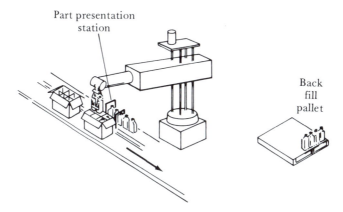

Part presentation
station

Back
fill
pallet

Figure 13–3   Robot Palletizing Bottles into a Carton

position for the manipulator to pick them up and load them into the box. The manipulator picks up two of the bottles and places them in the carton.

The manipulator repeats the cycle by picking up two more bottles from the parts presentation station and placing these two bottles in front of the two that have already been put in the box. On the final cycle, the manipulator picks up two more bottles and rotates them 90° in order to fill the final two open spots in the box.

When this task is finished, the controller gives a signal to the conveyor to index the table and present the next empty box to be filled. If the conveyor for the parts presentation station does not contain bottles, the manipulator will go to a subroutine to pick up bottles from the back fill pallet located in the work cell.

The flexibility of the palletizing routine is illustrated in Figure 13–4. In this application, the robot is operating in conjunction with two conveyor lines and is palletizing two separate pallets. Separate conveyors and pallets are needed because different parts are moved on each conveyor. The robot reaches over to conveyor one and lifts a part off the line. The robot then stacks the part onto the pallet directly in front of that conveyor line. Next, the robot reaches over to conveyor two and picks up a part from this line. The robot then stacks the part onto the pallet directly in front of that conveyor. In the routing for this operation, the controller sets an alarm condition when each pallet is full. This alarm signals the operator to come and remove the full pallet and replace it with an empty one. This entire operation could be accomplished by robots, though. A robot could be programmed to remove the full pallet and transport it to the next workstation.

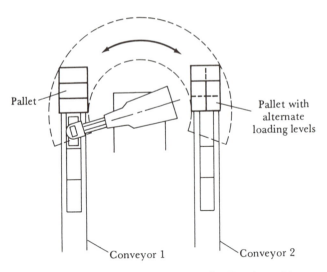

Figure 13–4   Palletizing Arrangement for Two Input Lines

Figure 13–5   Palletizing Routine Using an Articulate Coordinate Manipulator (Courtesy of GMFanuc Inc.)

Figure 13–5 illustrates a typical palletizing operation with an articulate coordinate manipulator. The specially designed gripper is used to pick up the parts as well as to load a spacer between the various rows of the pallet. In this operation, the manipulator grasps six parts at a time and places those parts on the pallet.

Palletizing with a robot can also be integrated with the work of a human. Figure 13–6 illustrates a robot palletizing air conditioner units in a box and a worker placing separators between the units in the box. In operation, the robot removes the air conditioner from the assembly line, places the unit in the box, and returns to the perch position to wait for the next air conditioner unit. The robot in this operation will begin its cycle every 19 seconds.

Line Tracking

Line tracking is a process in which the robot travels along with the part on an assembly line. This tracking might only be for a few

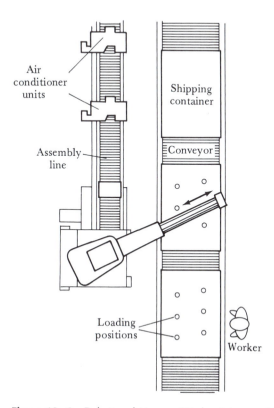

Figure 13–6   Robot and Human Worker Integrated for Assembly Line Tasks

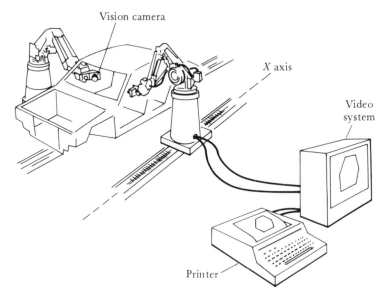

Figure 13-7   Line Tracking in an Inspection Process

feet. But while the robot is tracking the part, it can perform operations on the part.

   Figure 13-7 illustrates a typical line-tracking operation used for inspection of car bodies on an assembly line. Two articulate manipulators are used, one for each side of the car body. As the body comes to the inspection station, the manipulators enter the door opening of the car and inspect the critical welds on the body. The inspection process is accomplished through the use of a vision camera located on the end of the manipulator's arm. The vision system interfaces with a printer at the operator's workstation so that a printout of the inspection can be made. The operator monitors the inspection process in order to identify defects in welding.

   The manipulators in Figure 13-7 track the car body down the assembly line. Tracking is used in this application because so many cars must be produced each hour that the car body cannot be stopped for inspection. Therefore, the manipulator must move along with the body to accomplish the inspection process. The additional axis in this operation is the linear axis, or X axis. This axis is the track in which the manipulator moves in the linear direction.

Process Flow

All tasks in a material-handling operation must fit together for **process flow**. That is, the flow of the operation must move from one station to the next station without interruption.

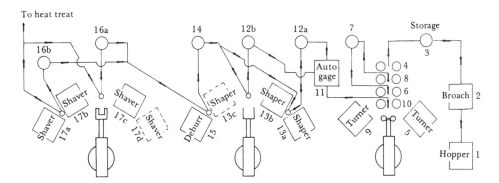

Figure 13–8   Sequence in Process Flow

Figure 13–8 illustrates a process flow for gear handling in a robotic system. The sequence of the flow is as follows:

1. Parts are delivered by a hi-lo and dumped into a hopper.
2. The parts are positioned for ID broaching by a push rod loader.
3. Parts are deposited for storage until the system is ready for these parts.
4. On command from the robot, the storage supply deposits two parts at pickup point 4.
5. The robot places the unfinished part into a turning operation. When the machine cycle is complete, the manipulator reaches into the turning operation and removes the part.
6. The manipulator places the part into station 6. The manipulator then returns to station 4, picks up two unfinished parts, and starts the cycle over again.
7. After leaving the temporary storage area, the robot places the part in a finished storage area.
8. Storage unit 7 supplies parts to location 8. The parts are presented to the robot for pickup two at a time.
9. The operation here is the same as the operation performed in steps 5 and 6.
10. The robot transfers parts from turning operation 9 to discharge table 10. From the discharge table, parts are transferred on a roller conveyor and a lift, which moves the parts to gaging station 11.
11. The gaging station checks the inside diameter and the outside diameter of the parts. Rejected parts are set aside in a holding chute. Accepted parts are transferred via the conveyor to an elevator.

12. The parts on the conveyor are placed in storage areas 12a and 12b.

13. At this location, the manipulator feeds three different shaping machines, 13a, 13b, and 13c. These machines shape the metal into gears and stamp the correct part number on each of the gears. While machine 13a is in cycle, the manipulator unloads machine 13b and transfers the shaped parts to a parts storage station at location 14. The manipulator then returns to the loading station at 12b, picks up an unfinished part, and loads that part into the 13c workstation. The cycle continues as the manipulator loads and unloads the various shaping machines.

14. After the gear-shaping process is complete, the robot picks up the part and deposits it in the storage area.

15. The parts are transferred via a robot to the deburring cycle. The robot places the gear in the deburring machine. Upon completion of the cycle, the robot reaches into the machine and turns the part over so that the other side is deburred.

16. The parts are placed on a conveyor line for storage at locations 16a and 16b.

17. The parts are shaved in the shaving machine. Four shaving areas are controlled by the robot, 17a, 17b, 17c, and 17d. Parts are then transferred by the robot to the heat-treating process.

MACHINE LOADING AND UNLOADING

The **loading** and the **unloading of machines** are other operations that are well suited for robotics applications. In fact, as was shown in Figure 13–8, loading and unloading parts is often combined with other robotic tasks in the manufacturing process. Typical manipulators used in machine loading and unloading are cylindrical coordinate manipulators and articulate manipulators. The main concerns in this application are the reach of the manipulator, the number of axes of the manipulator, and the weight capacity of the manipulator.

Figure 13–9 illustrates a machine-loading and -unloading application. In this application, the manipulator is a cylindrical coordinate manipulator. This type of manipulator provides the base rotation needed for reaching the indexing table and the lathe door, and it provides the reach for picking up and dropping off parts.

In this operation, the robot has complete control over the entire work cell. That is, the robot controls the indexing of the parts table, the start of the machine cycle, and the articulation of the manipulator. In the processing of this task, the robot controller provides an output signal through the interfacing lines to the indexing table.

Figure 13–9   Machine Loading and Unloading (Courtesy of GMFanuc Inc.)

This signal advances the table pallets one by one and presents an unfinished part for the manipulator to grasp. The manipulator places the unfinished part into the lathe. The robot controller then provides the lathe with a start signal. The lathe door is closed, and the machining operation is completed. When the lathe gives a finish signal, the manipulator moves into position and picks up the finished part from the lathe. The manipulator then places the finished part back onto the indexing table. The cycle is repeated for another unfinished part.

In the machine-loading and -unloading operation of Figure 13–9, the timing of the robot and the lathe must be coordinated. The time required for the robot or the machine to do its job is its cycle time. The robot's cycle time must match the lathe's cycle time for efficient operation. It is important that communication is established between the controller, the lathe, and the indexing table. This communication is handled through the robot's I/O.

DIE CASTING

An example of **die casting** is shown in Figure 13–10. In this operation, a wax pattern is dipped into a ceramic slurry and then dipped

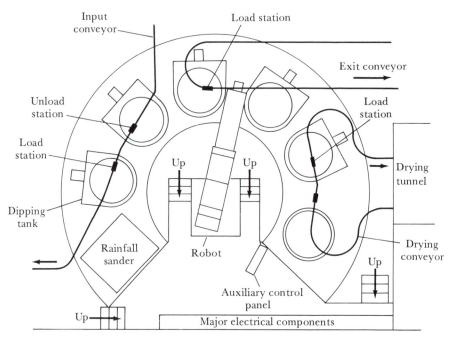

Figure 13–10   Die Casting

into a wet slurry of sand. This process may be repeated up to six times so that the wax pattern has the necessary thickness.

For this application, the robot must have certain characteristics. For example, the controller's memory must be large enough to store the various subprograms needed to move the manipulator to the different dip tanks. This memory size should be at least 120K bytes of memory. Also, the velocity of the manipulator's axis must be fast enough to meet the desired production of molds during a work shift. The work envelope of the robot must be large enough for the robot to reach several different dipping tanks. Finally, the **cell controller** must be able to control peripheral devices around the work envelope. In this example, the cell controller will be a **programmable logic controller.** The programmable logic controller will issue commands to the various components within the work cell commanding their operation. The programmable logic controller will also be used to generate reports, from detailing the amount of production to generating error reports to an operator.

The robot used in the application shown in Figure 13–10 must be able to handle at least 400 pounds. The robot controller must have a large memory capacity because there might be as many as 50 different die castings that will require different dipping times.

In this application, the robot reaches to the feeder line and picks a pattern from the conveyor. The pattern is transferred to the first predipping tank, where the robot controller controls the dipping time. Then, the robot removes the pattern to the next operation. The robot continues its cycle until the correct thickness of material has been deposited around the shell of the pattern. Upon completion of the dipping cycle, the robot returns the pattern to the conveyor line to the unloading station. The robot then picks a new pattern from the line and initiates another cycle.

In the example of Figure 13–10, the robotic system works in conjunction with the conveyor system. Parts are presented by the conveyor to the robot. The robot picks the patterns from the line and transfers the parts to the dipping tanks. Upon completion of the dipping process, the robot transfers the parts to the drying conveyor.

The total cycle time for this operation is between three and four minutes from start to finish of the cycle. Each of the steps of the casting process is timed to the exact sequence of the cycle. Again, the programmable logic controller is used here to time the sequence of events.

This robotic operation saves the company many dollars in waste and scrap, and it also removes the worker from a tedious, repetitive task. The savings result from the manipulator's control of the amount of time the mold is placed in the wax. This automatic control gives equal dipping time for each application of wax.

WELDING

The welding application is another area in which the robotic system is well used. Welding applications are broken down into two types: arc welding and spot welding. Each type is discussed in the subsections that follow.

Robots for welding applications must have several characteristics. The robot must be articulate, and many welding applications require a robot with a six-axis articulated manipulator. Also, the robot's axis speed must be sufficient to run the special paths required in the arc- and spot-welding applications. Finally, the robot controller must have the memory capacity to accomplish the welding task.

Arc Welding

Figure 13–11 illustrates a typical arc-welding work cell. Notice that a variety of components make up this work cell. The manipulator in this example is a five-axis articulate model with an arc-welding gun attached as end-of-arm tooling. The manipulator is mounted on a track that has a positioning table. The positioning table puts the part in the correct attitude in front of the manipulator. The positioning table has a dual function. That is, the operator can unload a finished part from the table and load an unfinished part on the table while the manipulator is arc-welding another part. The table then swings 180°, placing the unfinished part in front of the robot. A screen is placed between the operator and the manipulator to protect the operator from the flash generated by the arc-welding process.

The arc-welding controller is also shown in Figure 13–11. The controller supplies the necessary voltage, current, and wire feed for the arc-welding operation. The robot controller is situated behind the manipulator and is responsible for controlling the manipulator's movements. The control panel for the entire work cell is located directly next to the operator. From this location, the operator can control the positioning table, the arc-welding operation, and the manipulator's operation.

The arc-welding robot must have certain features. For instance, the robot must be able to adjust its axis speed during a program operation. The robot must have programs for cleaning the nozzle of the arc-welding unit. The robot must be able to adjust the voltage and amperage of the welding unit. The robot must be able

Figure 13–11   Arc-Welding Work Cell (Courtesy of Hobart Brothers Company)

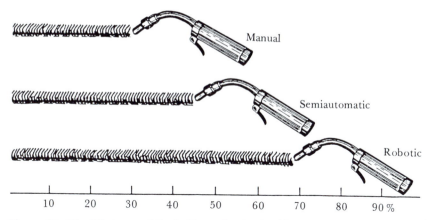

Figure 13–12   Efficiency of Cycle Times for Arc Welding

to change welding parameters during the operation. Finally, the robot must be able to meet production cycle times. Figure 13–12 illustrates the efficiency in meeting cycle times for robotic arc welding in comparison with semiautomatic and manual operations. The robot is about 70% efficient, while the manual operation is only about 30% efficient.

The robotic arc-welding cell may be fully automatic or semiautomatic. In the fully automatic system, there is no human intervention in the welding process. In the semiautomatic operation, a human works with the robotic arc welder.

Figure 13–13 illustrates a semiautomatic arc-welding application. The worker in this application loads the weld positioner. The robot then moves toward the weld positioner that has just been loaded and starts the welding routine. Meanwhile, the worker moves to the second positioner, unloads the positioner, and loads it with another part. (If this operation were fully automatic, a robot would also load and unload the positioner tables.) This process is repeated many times during normal operation. Thus, the arc-welding operation requires the robot to have repeatability. That is, the robot must be able to trace the arc-welding torch over and over in the same location.

Outside peripheral components are also important in the arc-welding operation. The robotic controller must be able to control the arc-welding equipment as well as the positioner tables. Thus, the controller must have enough input/output interfacing capabilities to meet these demands. Figure 13–13 illustrates the various types of interfacing signals that are required for the arc-welding operation. The controller controls the speed of the wire feed and the arc voltage. The arc voltage is directly related to the speed of

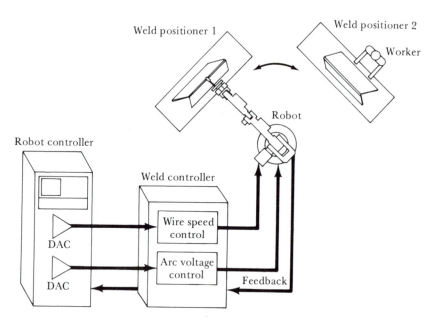

Figure 13–13 Semiautomatic Arc Welding

the wire feed. Therefore, the controller constantly monitors the arc voltage. Since the arc voltage is variable, it is an analog signal. In order for the controller to read this signal, it must be converted from analog to digital. This conversion is the job of the digital-to-analog converter (D/A) in the controller. If the arc voltage needs adjustment, the controller commands the D/A for the wire feed to increase or decrease the wire feed rate, which, in turn, adjusts the arc voltage. This constant monitoring of the circuit allows the robot controller to have complete control over the welding operation.

The robotic system's program should contain the different weaving patterns for the welding process. The standard weaving pattern can be adjusted for amplitude and frequency of the weave, as can other weaving patterns that might be used in a welding application.

The movement of the hand on which the welding torch is mounted must be controlled. Therefore, the robot must also have software that controls hand direction. This control keeps the robot's hand moving in a constant direction.

Another control for many arc-welding units is the seam-tracking function, which is illustrated in Figure 13–14. The **seam-tracking function** allows the end-of-arm tooling to locate and follow an actual weld joint. In Figure 13–14, the seam-tracking function senses the arc voltage. At the center of the weld, a certain voltage must

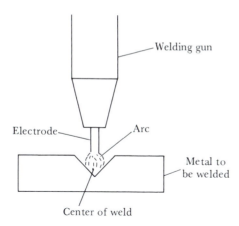

Figure 13–14 Seam Tracking

be generated to produce a good arc. If the welding unit starts to drift, the arc voltage changes. This change is sensed by the seam-tracking function, which sends a signal to the controller. The controller then automatically adjusts the manipulator's axis so that the manipulator is welding in the center of the seam again.

Spot Welding

The articulate robot for **spot welding** must have accuracy and repeatability. The robot controller must have adequate memory capacity to store all the different spot-welding programs. The robot must also have the weight capacity to carry the spot-welding gun, and it must control input/output circuits in order to interface with the weld controller. Finally, the robot must have the capacity for reach.

Figure 13–15 illustrates a typical spot-welding application on a production line. In this example, cars pass in front of the welding station. On each side of the line, two robots are mounted, one for the right-hand welds and one for the left-hand welds. Each of the robots is responsible for welds on a certain area of the car body. For example, the robots located at station 2 weld the inner door assembly. Once the weld has been completed, the car moves to the next station for the next set of welds.

In this operation, notice that the robots on the line are very close to each other's work envelope. Thus, special signals must be programmed into the controller to detect when one robot is in the other's work envelope. The signals used for this transfer are called **interference signals**. Each robot has a right and a left interference zone. In Figure 13–15, the right interference zone for the robot at

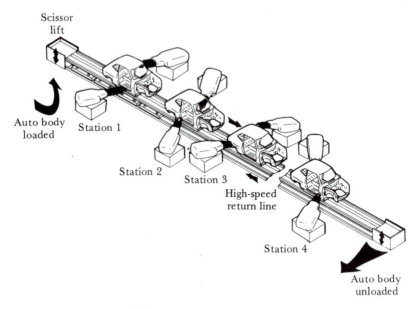

Scissor
lift

Auto body
loaded

Station 1

Station 2

Station 3

High-speed
return line

Station 4

Auto body
unloaded

Figure 13–15   Spot Welding

station 2 is the same as the left interference zone for the robot at station 3. Before robot 3 can operate within the left zone, robot 2 must be cleared out of the zone.

The controllers used with these robotic systems are high-technology controllers. These controllers allow the operator to constantly monitor the robot stations in order to identify the number of spot welds that have been accomplished in one shift. Also, these controllers have the large memory capacity needed for storage of the different weld paths for the different body styles. For example, for the application in Figure 13–15, the body style code is read at the staging station, which is located before station 1. As the car body waits to be transferred down the welding line, the body style code is fed to each robot. Once it has completed the weld for the body it is working on, the robot uploads the new style code and calls up the program for that body style. This process continues throughout the shift's normal operation.

The controller must also contain software that provides welding operations. For instance, in spot-welding applications, it is very important that the welding gun be clamped on the metal for a period of time. This period is called the **squeeze time**. Another important characteristic of spot welding is the amount of heat applied to the welding tips. A third concern is the **hold time**. Each of these signals can be programmed through the robot controller. The interfacing of these signals controls the spot-welding process.

The typical welding commands for spot-welding operations are as follows:

— Squeeze: Two tips are squeezed onto the part to be welded.
— Weld: Current is passed through the tips of the weld gun.
— Hold: For a specified period of time, the tips remain closed while the weld cools.

As the tips of the welding gun squeeze the metal, 800 to 1000 pounds of pressure are applied per square inch. The weld cycle provides current from the tips through the metal. This current flow generates the heat for the weld. The hold cycle keeps the tips on the weld spot during the cooling cycle of the tips.

Figure 13–16 illustrates a typical welding application.

Figure 13–16   Welding Application (Courtesy of Hobart Brothers Company)

INSPECTION

Inspection is a relatively new field for robotics application. The robot may have either an active or a passive role in the inspection. In the passive role, the robot feeds a gaging station with a part. While the gaging station is determining whether the part meets the specification, the robot waits for the process to finish. In the active role, the robot is responsible for determining whether the part is good or bad.

An application for an active robotic inspection process is illustrated in Figure 13–17. The robot in this example is programmed to trace the circuit trails on a printed circuit board. This inspection process uses both the computer of the vision system and the path control of the robot. The robot is programmed to identify defective signals from the vision computer and to remove defective printed circuit boards.

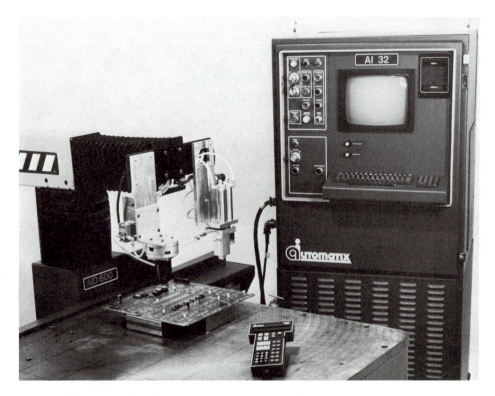

Figure 13–17   Robotic Inspection (Courtesy of Automatix Inc.)

ASSEMBLY

Assembly operations provide many applications for robots. Assembly applications are classified as batch assembly or low-volume assembly. In **batch assembly**, as many as 100,000 or 1,000,000 products might be assembled. These assembly operations have long production runs and require the same repetitive assembly routine. In **low-volume assembly**, a sample run of 10,000 or less products might be made.

The automation of the assembly area can be fixed automation or flexible automation. In **fixed automation**, the operation is repeated over and over. Fixed automation might be used in large-batch assembly. The fixed automation system cannot be easily reprogrammed when a new assembly operation is developed. In **flexible automation**, the assembly operation can be changed once the current assembly has been completed. Thus, once a low-volume production run is finished, the robot can be reprogrammed to perform another assembly operation.

The assembly robotic cell should be a modular cell. If the production run increases, more modular cells can be used. Figure 13–17 illustrates a typical assembly modular cell. Here, the manipulator assembles the components on the assembly line. This manipulator can be reprogrammed for another assembly operation once this job has been finished. Hence, this operation is flexible automation.

Another concern for the assembly robotic cell is the design of parts that can be assembled by the robot. That is, the robot's end effector must be able to easily pick up and assemble the parts. Minimizing the number of parts required in a robotic assembly is also critical.

Figure 13–18 illustrates some typical examples of assembly operations that can be changed for robotic assembly. In Figure 13–18A, the part contains several components that must be assembled. The part must be flipped over for the assembly, and several different moves are required for inserting the components. The new, preferred assembly is shown in Figure 13–18B. Here, the components and the part have been redesigned for robotic assembly. Notice that the part does not have to be flipped over now. All the components can be directly inserted into the main assembly block.

In Figure 13–18C, the part requires a large gripper shaft so that the gripper can grasp the part. Figure 13–18D shows the redesign of the part with a lip that allows a standard gripper to hold the part.

A common problem arising in assembly operations is illustrated in Figure 13–18E. Here, there are many different sizes of

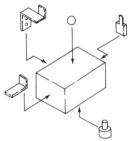

A. Difficult to assemble because of part orientation

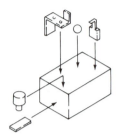

B. Preferred assembly because of new orientation

C. Difficult to grip because no lip provided

D. Preferred assembly because lip provided

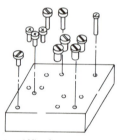

E. Difficult to automate because of too many screw variations

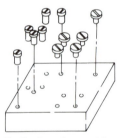

F. Preferred assembly because of fewer screw variations

Figure 13–18 Designing Parts for Robotic Assembly

screws that must be placed into the part. Each different-sized screw requires a different workstation for assembly. The design in Figure 13–18F solves the problem. The number of different screws has been reduced to two sizes.

PRINTED CIRCUIT BOARD ASSEMBLY

The expansion of robotics has reached into the manufacturing operations that require repeatability and accuracy during the assembly operation. One area that is well suited for robotics assembly is the insertion of odd electronic components. These components could be resistors, diodes, transistors, microprocessors, heat sinks, or a variety of other odd electronic components.

Figure 13–19 illustrates a typical overall electronic assembly operation. This work cell is defined as a flexible manufacturing electronic assembly cell. Notice that the total operation is made up of robots, conveyors, robotic hand changers, and human operators. The cell is identified as flexible because the system can generate a wide variety of manufactured printed circuit boards from one basic system structure.

Printed circuit boards are loaded onto carousels and rotated to be automatically placed on a transport mechanism, here identified as a conveyor, and then directed to work stations. A printed circuit board from the carousel is first read by a bar code reader. This bar code reader identifies the type of circuit board and alerts the system that a certain type of board is present in the system. A command will be issued throughout the entire system alerting the system stations to which components must be inserted into the board. The circuit board is then routed to the first station, which

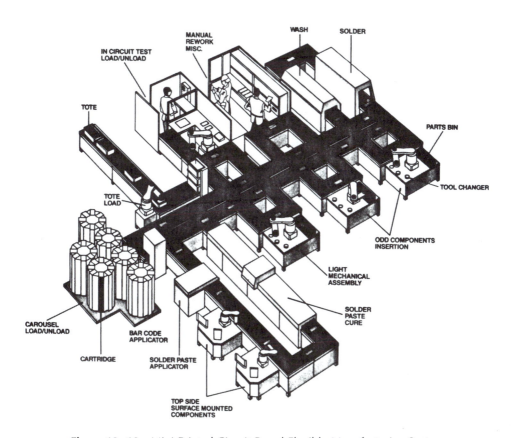

Figure 13–19   Mini Printed Circuit Board Flexible Manufacturing System

applies solder paste to it. The board is then transferred to two stations that insert the electronic components to the top side of the board.

After these two robotic stations complete their task, the board is routed to a station where the solder paste is heated. Once the operation is completed, a signal is sent from the robot controller to the cell controller. This signal identifies the operation as completed, and the cell controller indexes the printed circuit board to the next operation.

Next, the board is transferred to a light mechanical assembly station. If the printed circuit board requires this operation, then it is routed into the station. (This was identified earlier in the operation by the bar code reader.)

The printed circuit board is then transferred to the next station, an odd-component insertion station. Notice in this station that there are tool changers and an electronics parts storage bin. The tool changers are used so that the SCARA-type manipulator can work on one operation and then return the tool that had been used and pick up a new tool for another operation. This action reduces the need for additional manipulators but also, however, reduces the number of boards passed through the cell per hour.

The printed circuit board is then conveyed to a solder station, where the solder is flowed throughout the board. The board is then transferred to a wash station for cleaning. The next step to this operation occurs when the printed circuit board is conveyed to an inspection and rework station. Here, human interface with the system is employed. Technicians remove the printed circuit board from the conveyor and place it into an automatic test operation. The automatic test operation identifies whether the board is good or defective. Generally, the testers also identify which component is defective if the printed circuit board is rejected. If the board is defective, a technician will then remove the defective component and place the printed circuit board on the conveyor to be directed to the tote load station.

The tote load station is the location where printed circuit boards are removed from the conveyor and placed in totes. These totes are then transferred to the end of the conveyor for a removal-and-storage or shipping operation.

SPRAY PAINTING

One of the most common tasks for robotics application is **spray painting**. The robot can be used to apply paint to a variety of different products, such as automobiles, consumer products, and household goods. Because of this robotic application, the human worker no longer has to work in an environment that contains toxic paint fumes.

The robots used in this area are high-technology, articulate robots. The manipulators have at least six axes and, in some cases, have up to nine axes. These axes provide the manipulator with the correct axis movement to paint various areas of the product.

Figure 13–20 shows a typical spray-painting operation. This figure illustrates some of the concerns in developing a spray-painting application. First, the spray-painting operation must be done in some type of booth. The booth develops a constant airflow that takes the overspray and circulates it through a ventilation system. The ventilation system keeps the paint mist from getting all over the plant.

The second concern in a spray-painting application is the spray gun. The gun must be mounted to the manipulator's end effector in such a way that the articulation of the manipulator allows the gun to reach all the painting locations. The gun must also be easily cleaned so that the same robot can paint a variety of different colors.

Figure 13–20   Spray Painting (Courtesy of The DeVilbiss Company)

The final concern of the spray-painting application is the controller. The controller should have enough memory capacity to store all the different paths for painting the various products. The controller must also have enough input/output circuits so that it can control the color application, the hydraulic supply, and the various interfacings with plant computers.

The drive system for the spray-painting robots has been confined to hydraulic drives. Because of the explosive nature of paint fumes and the arc that might be created by an electric motor, hydraulic drives have been preferred. But new technology is currently being developed so that electric motors can be used in this type of environment.

Figure 13–21 illustrates another application for spray-painting manipulators. In this application, the manipulator is applying a polyurethane coating to a missile. The operation of this manipulator is about the same as the operation of a spray-painting manipulator. The main difference is the housing used to supply the polyurethane to the end effector. The housing must be flexible enough so that it can move the end effector around the unit.

Figure 13–21   Applying a Polyurethane Coat (Courtesy of The DeVilbiss Company)

≡ SUMMARY

Robots can be applied in many different areas in the workplace. Many of the dangerous and boring jobs can now be done by robots. The robot can be used in areas such as material handling, machine loading/unloading, die casting, welding, inspection, assembly, and spray painting.

One of the most common tasks for the manipulator in the industrial workplace is material handling. The manipulator used in this application must have the correct articulation to reach parts on a conveyor and must have the capacity to lift parts to overhead conveyor lines. The robot controller must have a large enough memory to store all the various routines that are required for material handling. The robot must also have the necessary axis speed to keep up with the assembly line in order to meet production schedules.

In material handling, two tasks are especially well suited for robots. In palletizing, the manipulator automatically stacks and unstacks parts from a pallet. In line tracking, the manipulator moves with the assembly line and performs its tasks.

Process flow of the manufacturing operation is also important. This flow enables the engineer to establish correct cycle times and the correct timing of the robot's tasks.

In machine loading and unloading, the robot loads unfinished parts into a machine. When the machining cycle is complete, the robot removes the finished part. The robots used in this area are either cylindrical coordinate or articulate robots. The main concerns are the manipulator's reach, its lifting capacity, and its speed.

In the die-casting application, a mold is dipped into wax solutions. The concern in this application is that the correct thickness of wax be deposited on the mold.

The largest area for robotics application is the welding operation. The welding operation can be either arc welding or spot welding. The manipulators used in spot welding are generally large articulator manipulators that have the capacity for lifting 80 kilograms or more. The arc-welding operation requires a small articulate manipulator that has high axis velocity.

Through programming and the use of sensors, the robot can inspect parts on an assembly line. The robot can use a vision sensor or limit switches to inspect the parts.

The assembly operation requires small cylindrical coordinate manipulators with very high axis velocities in order to meet production schedules. The assembly operation may be fixed automation or flexible automation. In some applications, the parts assembled by the robot must be redesigned for an efficient operation.

In spray-painting applications, the manipulators have hydraulic actuation systems because of the explosive fumes generated by the paint. The painting unit is placed inside a booth that provides ventilation and that removes paint mist.

KEY TERMS

arc welding
assembly
batch assembly
cell controller
die casting
fixed automation
flexible automation
hold time
inspection
interference signals
line tracking

low-volume assembly
machine loading/unloading
material handling
palletizing
programmable logic controller
process flow
seam-tracking function
spot welding
spray painting
squeeze time
welding

QUESTIONS

1. What is the best type of robot for material-handling operations?

2. Define the term *palletizing*.

3. In material handling, how many pallets can the manipulator palletize?

4. What condition signals the operator that a pallet is full?

5. Which axis is used for linear motion in line tracking?

6. In process flow, what devices are placed between the various press operations?

7. Which type of manipulator is best suited for machine loading and unloading?

8. Draw a work cell for a simple machine loading/unloading operation with parts input on the left of the manipulator, a milling machine, and parts output on the right of the manipulator.

9. What type of gripper allows machine loading and unloading to be done in one cycle?

10. What is a typical cycle time for a die-casting operation using a robotic system?

11. What are the two main welding areas in which robots are used?

12. What type of manipulator is generally used in welding applications, and how many axes does it have?

13. What is the most efficient arc-welding operation?

14. What equipment is used in arc-welding applications to present the part to the manipulator?

15. What two components of the welding weave pattern can be adjusted by the programmer?

16. What characteristic must an arc-welding robotic system have?

17. What are three characteristics of a spot-welding manipulator?

18. Since spot-welding manipulators must be very close to the production line, what safety features must be programmed into the robots?

19. What two sensors can be used for the inspection process?

20. For large-batch assembly, which type of automation is best?

21. What types of manipulators are best suited for assembly applications?

22. How many axes should a spray-painting manipulator have?

23. What type of actuators are found on spray-painting manipulators?

24. What is the purpose of the booth in the work cell of the spray-painting manipulator?

14

Robotics Applications: Growth and Costs

OBJECTIVES

Upon completing this chapter, you should be familiar with:

— Growth of robotics applications,
— Cost justification,
— Worker displacement,
— Training.

INTRODUCTION

Of the many application areas of robotics, the largest growth rate is in industry. Here, the robot is employed to displace workers that are involved in dehumanizing types of jobs. The robot is also used in areas where jobs are very routine and dull or where jobs are dangerous for the worker. When a plant uses robots, the workers can be removed from dehumanizing and dangerous jobs and be placed in jobs that give them personal satisfaction. Although the robots in these applications displace workers, the displaced worker can generally be transferred to other departments within the corporation and retrained for a new, perhaps more rewarding task.

A robot used in the manufacturing environment is classified as a capital investment for the corporation. Since it is a capital investment, many concerns must be explored before the new equipment is purchased and installed. These concerns can range from an increase in production to a reduced work force. These concerns constitute the return on investment for the corporation. If the return on the new investment is not high enough, then the chances are that the robot installation will not take place.

For all robotics applications, workers will have to be trained so that they can use and operate the robots. Training may be directed to individuals who are already in the work force, which is classified as retraining, or to individuals who are about to enter the field of robotics as a career.

This chapter will discuss these topics of growth rate of applications, cost justification, worker displacement, and worker training. As we will see, the future for robotics applications looks promising.

GROWTH OF ROBOTICS APPLICATIONS

Figure 14–1 gives a projection of the U.S. robot population for the year 1990. The various applications for the robot are shown on the horizontal axis of the graph. These applications are welding, assembly, painting, machine loading and unloading, and other applications. The estimates of the number of robots that might be used in these applications are shown on the vertical axis of the graph. Both a low and a high estimate are given.

Figure 14–2 is a table listing of projected robotics applications in one typical industry (automaker). This table lists the different applications of the robots and the projected growth rate for each application to the year 1990. In the following subsections, we will

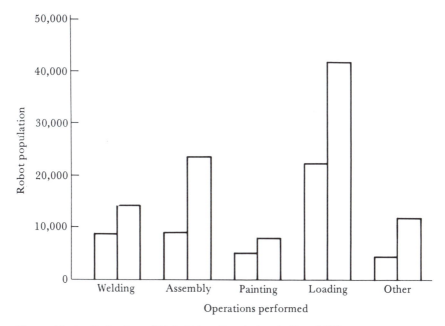

Figure 14–1 Projection of U.S. Robot Population in Year 1990

Application	Number of robots in use				
	1980	1983	1985	1988	1990
Arc and spot welding	138	1000	1700	2500	2700
Assembly	17	675	1200	3200	5000
Painting	47	300	650	1200	1500
Machine loading	68	200	1200	2600	4000
Parts transfer	32	125	250	500	800
Total	302	2300	5000	10,000	14,000

Figure 14–2 Projected Robot Applications for an Automobile Manufacturer

examine, in detail, the robotics applications and the projections listed in both Figures 14–1 and 14–2.

Welding

The first application illustrated in Figure 14–1 is welding. Welding is typically divided into two areas: spot welding and arc welding. In Figure 14–1, the low estimate for the welding application is about 9000 robots; the high estimate for this application is about 13,000 robots, industrywide, by the year 1990.

Moving to Figure 14–2, we see that, for the welding application, the automobile manufacturer used only 138 robots in 1980. This low number results from two factors. First, in 1980, the technology of the robot was not at the level necessary to meet the various production demands of the robot for this application. Second, the engineers who developed the applications were not familiar with the capabilities of robots.

However, according to Figure 14–2, the number of robots that will be used in the welding application is estimated to grow rapidly. By the year 1990, for example, the projected number of robots in the welding application is 2700 robots, just in this one industry. This growth potential is due to the advancement of robotics technology for the welding area and to the production engineers' greater understanding of how the robot can be used in different welding applications.

The big breakthrough in the arc-welding application will come when a vision system can be adapted to the manipulator. The addition of the vision system will allow the manipulator to track along a welding path and make path corrections. These welding path corrections can then be stored in the robot controller's memory for recall in the vision operation.

Robots that are used in these welding applications free the

worker from lifting the heavy spot-welding equipment employed to spot-weld car bodies. Also, the worker will no longer have to take the risk of inhaling dangerous fumes when welding aluminum and other materials that give off toxic fumes.

Assembly

The next application illustrated on Figure 14–1 is assembly. The assembly applications range from electronic assembly to the assembly of automobiles. This area has a very good potential for growth in the next decade. According to Figure 14–1, the number of robots used in this application is estimated to range from a low of about 9000 robots to a high of about 22,000 robots by the year 1990.

Figure 14–2 supports this growth potential for assembly robots. The automaker projects that, by the year 1990, 5000 robots will be used in assembly.

The greatest advance in this area will come when various sensors can be used in conjunction with the robot, sensors such as the touch sensor and the vision sensor. Also of importance in this area is the articulation of the manipulator. That is, the manipulator will have to simulate the human's coordination between vision and movement of the hand.

Painting

Painting is the next application that is illustrated in Figure 14–1. This application is one in which the robot has already been employed for many years. The painting application is broken down into spray painting and the dispensing of varnish and clear coats to the various parts found in a manufacturing environment. Notice that the number of robots projected to be used in the year 1990 is very low compared with the estimates for other applications. The main reason for the low estimate is that robots have been employed in this application for a long period of time.

In Figure 14–2, notice that the manufacturer is projecting that 1500 robots will be used for the painting application by the year 1990. Once again, the automaker's projections support a general growth rate for robot use in this application.

Machine Loading/Unloading

The next application on Figure 14–1 is machine loading/unloading. This application has the highest potential growth rate of any type of robotics application. The industry projects that at least 25,000 robots will be used in loading/unloading by the year 1990 and that as many as 45,000 robots may be used.

In Figure 14–2, this application is broken down into two parts: machine loading/unloading and parts transfer. Notice that, in 1980, the manufacturer used only 100 robots in these two loading applications. But the automaker projects that 4800 robots will be in use by 1990.

The largest growth for this application will come in the areas of parts being loaded and unloaded in various machining operations. Again, robot use in this area will increase as the technology of the robots increases. For example, the use of the vision sense and improvements in the robot controller and in the articulation of the robot will encourage the use of robots in the machine-loading application.

Other Applications

The final application listed in Figure 14–1 is called "other." This label includes all applications that do not fit into the welding, assembly, painting, or loading applications. For example, applications in this area might be robot use for applying sealant to car windshields or for deburring metal chips from a machined part. As illustrated in Figure 14–1, this area has a very low growth rate predicted for 1990.

The figure does not show all of the applications that will be filled by high-technology robot manipulators and controllers in the future. Many applications that are not even being thought of now may become quite popular in future years as the technology advances.

COST JUSTIFICATION

The cost of any robots that are to be used in an industrial area must be justified before the robots can be installed. This **cost justification** has three major components: expenditures, operating costs, and savings that can be generated from the robot installation. The data that is collected from these components can then be used in two calculations: the simple payback formula, which determines the number of years required to recover the expense of the robot installation, and the formula for the return on the investment of the corporation's money.

Application engineers must understand these basic components so that they can intelligently present their case for robot installation to upper management. In the following subsections, we will examine each component in detail.

Expenditures

The **expenditures** for robot installation are divided into five basic areas:

1. The cost of the robot,
2. The costs of support equipment,
3. The cost of installation,
4. The resale value of old equipment,
5. Investment tax credits.

The many considerations that must be taken into account in each area are discussed in the following paragraphs.

The **cost of the robot** includes the following items:

— The base cost of the robot;
— The cost of optional equipment for the robot, such as larger memory capacity, special panels for the robot controller, teach pendants, or special cables used for interfacing the controller and the manipulator;
— The cost of test equipment that is required for maintaining the manipulator and the controller;
— The cost of end-of-arm tooling that is required for the operation of the robot in the work cell;
— The cost of training support personnel;
— Application engineering costs;
— The cost of spare parts packages for robot maintenance.

The next area of expenditures that should be considered by the application engineer is the **cost of support equipment** that will be needed in the development of the robotic work cell. Some of the support components that might be needed are as follows:

— Conveyors to deliver parts to the robot's workstation;
— Safety or hazard enclosures around the work envelope of the manipulator;
— Automatic fixturing for the robotic operations;
— Additional equipment resulting from modifications to the work cell after installation.

The third area of expenditures for the robotic work cell involves the **installation costs**. These costs include the costs for preparing the robot site as well as for installing the robot. The following items may contribute to installation costs:

— Site preparation, which may include a new foundation for the floor, structural support for peripheral equipment, or enlargement of the space in which the robot will reside;
— Utility drops and connections for the robot and the peripheral equipment;
— Interface connections of pneumatic and electric lines to the peripheral equipment;
— Robot placement;
— Peripheral equipment placement;
— Old equipment removal and storage.

The fourth item of expenditures is a possible return of capital to the corporation. This return comes from the money that may be received from the **resale of old equipment** that the robot is replacing. The final item related to the expenditure side of robot installation is the **investment tax credit** that is given to the corporation for investment in new equipment. This credit can be as high as 10% of the total expenses that the corporation incurs for the robot installation.

Figure 14–3 presents a sample work sheet for the expenditure side of the cost justification for robot installation. Most of the costs and credits just described are included in this cost sheet.

Savings and Operating Costs

In this section, we will consider the second and third areas of cost justification of robot installation: the cost savings for the corporation and the operating costs after the robot is installed.

The **cost savings** include the following factors:

— Direct savings generated by the robot installation, such as savings in wages or benefits;
— Indirect savings per year resulting from less use of floor space, heat, lights, or other environmental items that are no longer needed because of the robot installation;
— Depreciation costs of the robot over a 10-year period, which are generally given by the depreciation rate for capital equipment installed in the manufacturing environment.

The **operating costs** that are a result of the robot installation are similar to the costs involved for any equipment installed in the manufacturing area. These costs might include the following:

— Maintenance cost for the robot per year, which is generally the labor cost associated with preventive maintenance and

```
                    EXPENDITURES


ROBOT
    1.   BASE COST ...................... $_____
    2.   OPTIONAL EQUIPMENT ............. $_____
    3.   SPECIAL TOOLS .................. $_____
    4.   MAINTENANCE AND TEST EQUIPMENT  $_____
    5.   ACCESSORIES .................... $_____
    6.   END-OF-ARM-TOOLING ............. $_____

             ROBOT SUBTOTAL .......   $_____

SUPPORT EQUIPMENT
    1.   CONVEYORS ...................... $_____
    2.   SAFETY ENCLOSURES .............. $_____
    3.   AUTOMATIC FIXTURING ............ $_____
    4.   OTHER MODIFICATIONS ............ $_____

         SUPPORT EQUIPMENT SUBTOTAL .... $_____

INSTALLATION COSTS
    1.   SITE PREPARATION ............... $_____
    2.   UTILITY CONNECTIONS ............ $_____
    3.   INTERFACE CONNECTION ........... $_____
    4.   ROBOT PLACEMENT ................ $_____
    5.   PERIPHERAL EQUIPMENT PLACEMENT   $_____
    6.   EQUIPMENT RELOCATION ........... $_____

         INSTALLATION COST SUBTOTAL .... $_____

OLD EQUIPMENT REMOVAL AND STORAGE COSTS $_____

         NET INSTALLATION COSTS ........ $_____

INVESTMENT TAX CREDIT ................. ($_____)

ADJUSTED INSTALLATION COSTS .......... $_____
                                       ================
```

Figure 14–3   Expenditure Cost Sheet

the major maintenance on the controller and the manipulator;
— Cost of the supplies required for the maintenance of the controller and the manipulator, which may also include the cost for storage of the spare parts;

```
┌─────────────────────────────────────────────────────────────────┐
│                  SAVINGS AND OPERATING COSTS                      │
│                                                                   │
│  OPERATING SAVINGS                                                │
│     1.   DIRECT SAVINGS PER YEAR ................ $_____ │
│     2.   INDIRECT SAVINGS PER YEAR .............. $_____ │
│     3.   DEPRECIATION ........................... $_____ │
│                                                                   │
│              ANNUAL SAVINGS SUBTOTAL .... $_____         │
│                                                                   │
│  OPERATING COSTS                                                  │
│     4.   MAINTENANCE COST PER YEAR (LABOR) ..... $_____  │
│     5.   MAINTENANCE AND REPAIRS PER YEAR ...... $_____  │
│     6.   OPERATING SUPPLIES PER YEAR ........... $_____  │
│     7.   TAXES AND INSURANCE ................... $_____  │
│     8.   OTHER COSTS PER YEAR .................. $_____  │
│                                                                   │
│              ANNUAL COSTS SUBTOTAL ........... $_____    │
│                                                                   │
│  NET ANNUAL SAVINGS BEFORE TAXES ............ $_____     │
│                                                                   │
│  TOTAL NET ANNUAL SAVINGS ................... $_____     │
│                                                                   │
└─────────────────────────────────────────────────────────────────┘
```

Figure 14–4   Savings and Operating Costs Work Sheet

— Taxes and insurance costs for the robot installation;
— Miscellaneous costs for the robot.

These various costs and savings for the robot installation are before-tax costs and savings. The taxes that are paid by the corporation must be taken into account before the total net annual savings can be stated for the robot installation.

Figure 14–4 presents a sample work sheet that can be used to determine the various operating costs and savings for the robot installation. Note that the work sheets of Figures 14–3 and 14–4 can be combined into one form that application engineers can use in developing the total cost picture for the robotic system.

Payback Period

The next step in cost justification of the robotic system is to determine the payback period. The **payback period** is the time required for the corporation to recover its initial investment. The simple formula for the payback period is as follows:

$$\text{payback period} = \frac{\text{AIC}}{\text{NAS}} \tag{14-1}$$

where

payback period = number of years before positive cash flow
is developed from the robotic installation

AIC = adjusted installation cost, which is generally taken from
the work sheet of Figure 14–3 and is the summary of
all of the expenditures for the robot installation

NAS = net annual savings, which is the total savings generated from the robot installation and is taken from
the work sheet of Figure 14–4

From this equation, the time required for the company to receive a positive cash flow from the installation can be calculated. A sample calculation for the payback period is illustrated in the following example.

Example
Determine the payback period for a robotic system that has a total net annual savings of $72,049.00 and an adjusted installation cost of $104,000.00.

Solution
Using Equation 14–1, we can calculate the payback period as follows:

$$\text{payback period} = \frac{\text{AIC}}{\text{NAS}} = \frac{\$104,000}{\$72,049} = 1.44 \text{ years}$$

In this example, the corporation will receive a return on its investment in the robotic system 1.44 years after the system is installed and begins production.

Return on Investment

The final cost justification item that an application engineer must consider is the **return on investment (ROI)**. The ROI figure gives the corporation a way to compare the return on the investment with the capital expenditure and determine if the capital meets or exceeds the cost of the project.

The formula for calculating the ROI is given next. The values for the terms in the equation are taken from the work sheets of Figures 14–3 and 14–4. The equation is as follows:

$$\text{ROI} = \frac{\text{AYR}}{\text{AIC}} \times 100 \qquad\qquad (14\text{--}2)$$

where

    ROI = return on investment, a figure that is based on a five-year plan

    AYR = average yearly return, which is the net savings and is taken from the work sheet of Figure 14–4

    AIC = adjusted installation cost, which is taken from the work sheet of Figure 14–3

A sample calculation for ROI is shown in the next example.

Example
Calculate the return on investment of a robotic installation that has an adjusted installation cost of $104,000.00 and an average yearly return of $72,049.00

Solution
Using Equation 14–2, we can determine the ROI as follows:

$$\text{ROI} = \frac{\text{AYR}}{\text{AIC}} \times 100 = \frac{72,049}{104,000} \times 100$$
$$= 0.693 \times 100 = 69.3\%$$

In this example, the return on investment is 69.3%, which is a favorable ROI for the robotic system. A typical ROI for a corporation investing in capital equipment is 20%.

WORKER DISPLACEMENT

As new robotics technology is used in the manufacturing area, the manual operation of jobs will be discontinued. Therefore, employees may be laid off or, in most cases, may be displaced. The term *displaced* means that the worker is transferred to another job within the manufacturing facility. The term *displaced* is used to indicate that the worker is not eliminated.

    Figure 14–5 illustrates a typical displacement chart. It gives the total displacement of workers because of automation installed in manufacturing operations in the United States. The displacement chart starts for the year 1979 and continues through the year 1990. The chart clearly shows that the displacement of workers will have a linear rise through the year 1990.

    Figure 14–6 is a table that gives estimates of worker displacement for the state of Michigan in 1990. Since Michigan is primarily

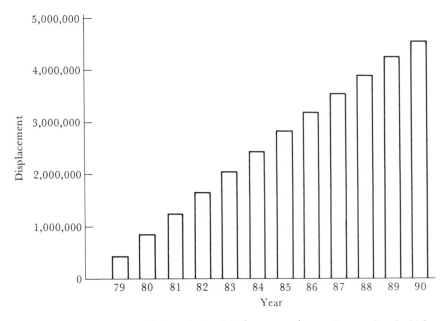

Figure 14–5   Projection of Cumulative Displacement due to Automation in U.S. Manufacturing Employment, 1979–1990

Application	Autos Range of estimate		Other manufacturing Range of estimate		Total Range of estimate	
	Low	High	Low	High	Low	High
Welding	2272	2912	450	820	2722	3732
Assembly	2984	6250	410	1228	3394	7478
Painting	1278	1776	262	450	1540	2226
Machine loading/unloading	3552	5682	1434	2786	4986	8468
Other	568	1138	312	860	880	1998
Total	10,654	17,758	2868	6144	13,522	23,902

Figure 14–6   Estimate of Job Displacement in Michigan by Robotic Application for 1990

a manufacturing state, the number of workers that will be displaced by robot installation is high. The number of workers that will be displaced is tabulated by the application of the robot. The industry areas are divided into automotive and all other manufacturing facilities. Notice that the machine loading/unloading area is expected to displace the greatest number of workers in automotive manufacturing. The next highest area of displaced workers is in the assembly operation. These figures support the projections of robot applications that were given earlier in Figure 14–1. The table also gives the total estimated number of displaced workers in Michigan for the year 1990 arising from the implementation of robotics. The total ranges from a low of 13,522 to a high of 23,902.

We can also consider worker displacement by age group. First, let's consider the 25–35-year-old group. This group of workers consists of individuals who have not as yet entered an apprenticeship program or have just started work in a skilled-trade area. When robots are installed on the assembly line, these individuals are the ones who are most likely to be displaced. These younger workers may have to return to school to learn about the new technology if they wish to remain in the work force in the future.

A second group of workers consists of individuals in the 35–45-year-old range. These individuals have developed a skill in their trade and are now journeymen. These workers will not be displaced because of the installation of the robot, but they may have to return to school to learn a new skill, such as programming, installing, or servicing of robots. Returning to school could pose a problem for workers in this age group. Probably, these people have not been in a classroom for a long time, and the idea of returning to school could be a little frightening.

A final group of workers affected by a robot installation consists of individuals in the 45–55-year-old range. These workers have been in their trade or on the assembly line for many years. Because of their seniority, these workers are the least likely to lose their jobs. And because of their experience, these workers may be asked to supervise the installation of the new robots. Many individuals in this age group are leaders in the plant. These older workers have seen many different types of automation and productivity improvements introduced into the plant over the years. For this reason, they will be responsible for seeing that the younger workers accept and become familiar with the new technology.

Workers in this age group that are close to the age of 55 may be displaced. These individuals are close to retirement age, and the company may ask them to take an early retirement. If these individuals elect to stay in the work force, they will have to return to school for retraining and upgrading of their skills to meet the demands of the new technology.

TRAINING

When a new technology appears in the manufacturing area, one of the most important concerns is how industry will find the individuals who can operate and service the new equipment. Here is where training comes in. In the robotics field, there are basically two types of training that will be required: the training offered by private technical schools or community colleges and the training offered by manufacturers of the robots, also known as vendor training. In this section, we examine both types of training.

Community College or Private School Training

The community colleges and the private schools must offer training at two levels. First, they must train workers who are about to enter the work force for the first time or who are returning to school to learn a new skill or trade. Second, they must train individuals who require prerequisite training.

New workers who are about to enter the work force for the first time in the area of robotics must acquire skills that match the skills demanded by the robotics industry. These skills are in the areas of basic robotic fundamentals, robotic programming, computer programming, fundamentals of electronics and electricity, fundamentals of microprocessors, digital electronics, tool design and fixturing, and blueprint reading. In addition, these individuals require several basic courses in the sciences and the arts.

Workers who need to be retrained or to have their skills upgraded to meet the demands of the new technology must be allowed to enroll in school, be retrained, and return to work as soon as possible. Figure 14–7 illustrates a triangle of learning that can be used for the worker needing skill upgrading. The bottom block is labeled prescriptive testing. Prescriptive testing is used by the school to determine the individual's weaknesses and strengths. From this testing procedure, the school can identify a prescriptive learning

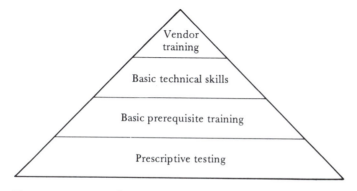

Figure 14–7 Triangle of Learning

path for the student. For example, this test could show that the student requires skill upgrading in basic mathematics, basic reading, basic English, or basic technical areas.

The next block in Figure 14–7 shows that basic prerequisite training is needed by the individual. This training could be in the area of English skills or technical skills. The next block shows that technical training is required for the individual. This training should be in the areas of electronics, CAD/CAM, computer programming, tool and die fixturing, and basic robotics.

The final block of the learning triangle is labeled vendor training. Vendor training is offered by the robot manufacturer to train customers that have purchased the robots for their manufacturing facility. Vendor training is described in the next subsection.

Vendor Training

Vendor training is training provided by the supplier of the robotic system. This training is very concentrated. It generally covers two areas of the robotic system: operation and programming. The operation and programming class takes the participant through a step-by-step procedure for operating and programming the robot. These courses discuss the various programming structures and the various codes that are necessary for robot operation in production.

Another course offered by the vendor is maintenance training. This course generally covers two basic areas: electrical troubleshooting of the robot controller and of the manipulator and mechanical operation and maintenance of the manipulator. These vendor courses are generally basic overviews. Advanced training courses in programming and maintenance of the robotic system may also be offered by the vendor. The advanced courses are generally one-week courses given at the vendor's training center or at the customer's location.

Many vendors provide additional training in special areas for their customers. For example, vendors may offer application courses. Application courses are generally geared for first-time users of robots. These courses give the user ideas for the efficient use of the robot as well as applications that have already been developed by the robot manufacturer. Another type of training that may be provided by the vendor is operator training. This training is designed for the operators who will be responsible for correcting simple problems in the robot's daily operation, calling up the various operating programs from the robot's memory, and ensuring that daily production schedules are met. Another vendor course that may be offered is mechanical teardown. This course is a maintenance training course that covers the internal mechanical operation of the manipulator and the replacement and repair of the manipulator. These special classes offered by the vendor may be conducted at the customer's facility or at the vendor's training center.

The people conducting the vendor's training sessions must assume that the individual attending the training has all the prerequisite skills necessary for the concentrated work. Thus, the customer should select individuals who meet the basic prerequisite skills for robotic training. If the customer's personnel do not have the necessary prerequisites, then the customer should contact the local private trade school or technical community college. These schools can provide the customer's personnel with the necessary prescriptive learning so that they can then take advantage of the vendor's training.

SUMMARY

The major applications projected for robots by 1990 are welding, assembly, painting, and machine loading/unloading. By 1990, the number of robots used in welding applications will have grown substantially. The growth in this application will increase when the robot's vision sense is able to track the welding path and make the necessary corrections during the welding process. In the assembly area, robot use is also going to expand by 1990. Growth in this application will further increase when the vision systems for the robot become more advanced, allowing the robot to perform the tasks of electronic assembly and mechanical assembly.

Robot use in painting applications shows the least growth, since robots have been used in this area for many years. Machine loading/unloading applications show the greatest growth. Use of robots in this area will expand as improvements are made in the robot controller and in the robot's articulation.

Application engineers must develop a cost justification for a planned robot installation. The cost justification work sheet includes expenditures, operating costs, savings, payback period, and return on investment (ROI).

The expenditures include the cost of the robot, the cost of support equipment, installation costs, the resale value of old equipment that the robot is replacing, and investment tax credits for the new equipment. The savings that will be generated from the robot installation include direct savings, indirect savings, and depreciation costs for the new equipment. The operating costs include the maintenance costs of the robotic system, supplies required for operation of the controller and the manipulator, taxes and insurance costs for the year, and miscellaneous costs for robot operation.

These costs and savings are used in a payback period formula that determines the time required for the company to recover the investment in the robotic system. This formula compares the adjusted installation costs with the net annual savings for the robotic system. Costs and savings are also used in the calculation of return on investment. This formula is a comparison between the average yearly return on the investment and the adjusted installation costs.

When new technology is introduced into an industry, some workers will be displaced from their jobs. Displacement means that the worker will not be eliminated but will be transferred to another position within the company and retrained for that new position.

Training is an essential part of the installation of a robotic system. Two types of training are identified: prerequisite training supplied by the local community college or private school and the vendor training supplied by the robot manufacturer.

Before the installation of the robotic system, the company should select the individuals who will be responsible for the robotic system and upgrade their skills as required. The skill upgrading can be accomplished through the use of prescriptive learning tests. These tests will identify which skill areas must be upgraded. This training should be completed before the personnel attend vendor training.

The vendor training schools generally offer courses in the operation and the programming of the robotic system and in the maintenance of the robotic system. Special courses may also be offered by the vendor; these courses may include applications courses, operator training courses, and mechanical teardown courses.

KEY TERMS

cost justification
cost of the robot
cost of support equipment
cost savings
expenditures
installation costs

investment tax credit
operating costs
payback period
resale of old equipment
return on investment
vendor training

QUESTIONS

1. What application of robotics will have the highest population of robots in the year 1990?

2. What application of robotics will have the lowest population of robots by the year 1990?

3. What two areas is the application of welding broken down into?

4. What robotic application will have the largest percentage growth rate by the year 1990?

5. What breakthrough in robotics technology will expand the welding application?

6. What are two painting applications for robots in a manufacturing facility?

7. What is the projected use of robots in machine loading/unloading by the year 1990?

8. Name two applications that can be placed into an "other applications" category.

In questions 9–14, identify the item as an expenditure, E, an operating cost, OP, or a savings, S.

9. Maintenance costs of the robot and peripheral equipment.

10. Costs of floor space, heat, lights, and ventilation.

11. Utility drops, peripheral equipment replacement, and site preparation.

12. End-of-arm tooling.

13. Old equipment removal and storage.

14. Conveyor removal and cost of new conveyor.

15. What is the payback period for a $79,750 robotic system that generates an annual savings of $23,250?

16. Which two factors does the return on investment compare for a robotic installation?

17. Which robotic application will have the largest worker displacement for the state of Michigan in the year 1990?

18. Which robotic application will displace the greatest number of workers in manufacturing facilities in the state of Michigan in 1990?

19. Define *worker displacement*.

20. Who is responsible for prerequisite training in the field of robotics?

21. What type of training should be supplied by the vendor?

22. What is the first step in training displaced workers?

23. Name three courses that should be available to a person wishing to upgrade skills in robotics.

24. Name three technical courses that should be offered to a person wishing to upgrade skills in robotics.

25. Name two courses that may be offered by the vendor of the robotic system.

15

Communication

OBJECTIVES

Upon completing this chapter, you should be familiar with:
— Data communication,
— Communication networks,
— Data transmission,
— RS–232 standard,
— Transmission modes,
— Networking the communication,
— Network cables,
— Network-to-network communication.

INTRODUCTION

With the implementation of automation on the plant floor, a need for mechanical and electrical devices to communicate with one another has come about. Communication allows these devices to transmit and receive special coded signals that enable the operator of a robotic system to query the system about its daily production or identify the number of parts produced within a given period of time. The communication is in the form of digital-encoded signals arranged in a certain pattern that enables the communication to be established. Often set up from the host computer of the plant to the peripheral devices within the robot's work cell, the communication developed is two-way; the devices within the cell can talk with the computer, and the computer can talk with the devices within the cell. This chapter focuses on the communication established between the various components within the work cell of a robot.

DATA COMMUNICATION

Data communication can take the form of any one of several methods used to communicate data from one location to another. One method of data communication, for example, is voice communication between two individuals. Data is sent from the voice of the sender to the ear of the receiver. The received data is then translated by the brain into phrases and words that are understood by the receiver.

In computer communication, these words and phrases must be sent through a different method so that the receiving end can understand them. The format that is established so that the receiving and the sending end can understand each other is called *protocol*. Figure 15–1 illustrates this type of data communication used with computer-controlled devices.

In Figure 15–1A, two points of digital communication are established between the computer and the printer. Notice that these

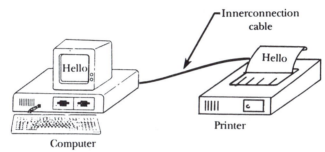

A. Digital communication between computer and printer

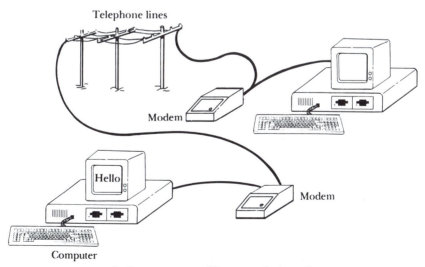

B. Two computers talking over telephone lines

Figure 15–1 Data Communication Used with Computer-Controlled Devices

two communicating devices are connected together through a cable. The communication developed is one-way; data is sent from the computer to the printer over a very short distance.

In Figure 15–1B, two computers are talking to each other over telephone lines. Since the distance between the two computers is great, special computer communication devices (modems) are required to make communication possible.

The communication illustrated in Figure 15–1 is computer or data communication in which information is being transmitted between two different points. These points are the *receiver location* and the *transmitting location*. The method that is used to communicate between these locations might seem very simple, but the complexity of sending and receiving correct information is very important to the robotic work cell.

Digital Signal

A **digital signal** is the common signal that is used to communicate between different computer-controlled devices. A digital signal is used because it is in one of two states, either on or off. Figure 15–2 illustrates a typical digital signal used in computer-controlled communication.

Notice in the figure that the signal starts from a zero reference point. From the zero reference point, the signal rises very rapidly to a level above the zero reference. These two points for the digital signal are very important to understand. The zero reference point is identified as a **logic 0** or OFF condition, while the level above the OFF condition is identified as a **logic 1** or ON condition. These two conditions in the digital signal mean that the signal is either on or off. This on-or-off state makes the digital signal very useful in computer circuitry.

The two states of the digital signal are also very important in computer communications. When put together, the digital signal levels form words of information that different computer-controlled devices can understand. The individual logic 1 and logic 0 are identified as **bits** of information. Figure 15–3A illustrates these bits of information. Figure 15–3B illustrates the combination of the bits to form a **byte** of information. Notice that one byte of information contains 8 bits. It is the combination of bytes that then makes up words understandable by computer-controlled devices.

Figure 15–2   Digital Signal Used in Computer Communication

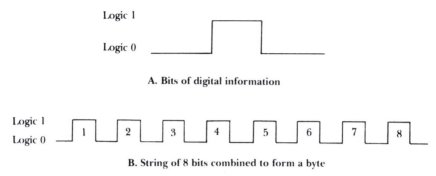

A. Bits of digital information

B. String of 8 bits combined to form a byte

Figure 15–3   Digital Signals

Digital Code

The formation of various bytes into a special arrangement makes up the codes that computer-controlled devices understand. The arrangement of bits into a special pattern that corresponds to the letters and symbols found on the keyboard of a computer has been standardized by the American Standard Code for Information Interchange, or **ASCII.** The ASCII code has a total of 128 characters, which include uppercase and lowercase letters, the numbers 0–9, and punctuation marks. Figure 15–4A shows a byte of the ASCII code used for the letter A; Figure 15–4B shows the ASCII code for the letter F, and Figure 15–4C, the ASCII code for the question mark.

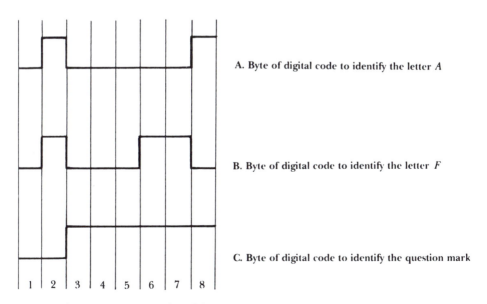

A. Byte of digital code to identify the letter A

B. Byte of digital code to identify the letter F

C. Byte of digital code to identify the question mark

Figure 15–4   Examples of the ASCII Code

COMMUNICATION NETWORK

A **communication network** is made up of many different components in order to establish communication. It can be thought of as a seven-step control to establish the communication.

Figure 15–5 illustrates a seven-stage communication network. The first stage, the component at point A, is identified as *data terminal equipment* (DTE). DTE refers to any telecommunication equipment such as mainframe computers, minicomputers, or robot controllers. The second stage of the communication network is the interface between the DTE and *data communication equipment* (DCE). This interface is connected in either a serial or a parallel mode. The device identified as the DCE is called a *modem*—short for modulator/demodulator. The modem (third stage) is responsible for modulating the signal so that it can be transmitted over the transmission channel (fourth stage) between point A and point B. At the receiving point, another DCE (fifth stage) is required to demodulate the signal so it can be read by the DTE, and another interface (sixth stage) is connected between the DCE and DTE. DTE equipment is the final point (seventh stage) of the communication.

Several different types of modems are used in computer-controlled communication. *Short-haul modems* link short distances together for communication purposes, while *long-haul modems* connect great distances. The connection of a robotic work cell to a host computer within the plant over a distance of 500 to 1000 feet or longer is one example of a long-haul modem. Another example is the connection of a robotic work cell in New York with a repair center in Detroit, with the communication established over tele-

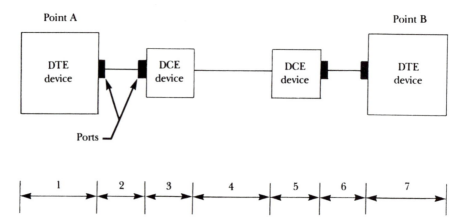

Figure 15–5 Seven-Stage Communication Network

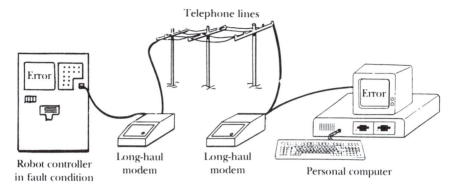

Figure 15–6 Long-Haul Modems Used to Transmit Errors over Distances

phone lines. Figure 15–6 illustrates the use of long-haul modems to transmit errors over distances.

DATA TRANSMISSION

In order for computer-controlled devices to speak to one another, they must speak the same language. This is accomplished through the use of the ASCII code. The devices must also transmit and receive data at the same speed and in a serial or a parallel mode. If any of these standards are not met, then the devices cannot communicate with one another.

Bit Rate

Data communication is measured in the number of bits that can be sent in one second. **Bit rate** is thus measured in bits per second (bps). Typically, bit rates range between 110 and 19,200 bps. The shorter the communication distances, the higher the bps rate will be. In the use of long-haul transmission, the bps rate ranges from 110 to 1200 bps; shorter-distance transmission can operate at a rate of 19,200 bps.

Baud Rate

Bit rate is sometimes confused with baud rate, but these two terms are not synonymous. **Baud rate** is defined as the number of times a signal changes in a given period of time. Figure 15–7 shows how baud rate is related to bit rate. Notice that the signal has three different states in a period of time. These three different states

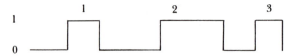

Figure 15–7   Baud Rate

mean that the signal changes from a logic 1 state to a logic 0 state. The rate of change is identified as the baud rate.

Serial Transmission

The bit-by-bit transmission of data is identified as **serial transmission.** Figure 15–8 illustrates this method of transmission of data. Notice that the first bit sent out is a *start bit,* which is followed by a byte of information and then by a *stop bit.* The data transmitted will follow in a sequential order until all of the information has been received.

The serial transmission mode is very popular because it requires only one wire to transmit the data from transmitter to receiver. It is very important, however, to understand that devices that communicate serially can communicate only with other devices that communicate serially.

Parallel Transmission

The method of transmitting a byte of information at the same time as a clock pulse is called **parallel transmission.** Figure 15–9 illustrates the method used to develop parallel communication. Notice that each bit of information contained in the byte has a separate line over which to communicate. A ninth wire is used as the **strobe** or clocking signal. All 8 bits of information will be received based on the clocking pulse. Notice that, as the clocking or strobe pulse is falling, the signals are sent. This type of transmission is *synchronous* (in step with the strobe). If the signals are sent without the coordination of the strobe pulse, then the signals are *asynchronous.*

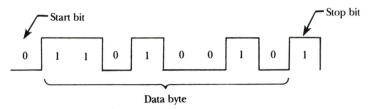

Figure 15–8   Serial Transmission with Start and Stop Bits

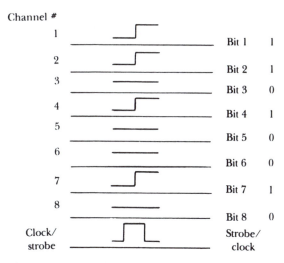

Figure 15–9 Parallel Transmission Using a Clock Pulse

One disadvantage of the parallel transmission mode is that 9 conductors are required to transmit a signal, as opposed to serial transmission, which requires only one wire.

RS–232 STANDARD

One of the mediums that is used to connect various computer-controlled devices together is cable. The cables that are employed require that a connector be attached to the transmitting end and also to the receiving end. So that all communication devices are connected in the same manner, a standard was developed. This standard is known as the *RS–232 standard interface*. The term *RS–232* comes from the 232 Recommended Standard.

The RS–232 standard is specially designed to meet the requirements of serial transmission. These requirements can be identified as follows:

— Specific pin assignments for ground, data, and control signals,
— Maximum cable length of 50 feet or 15 meters for transmission of signals,
— Maximum data transmission rate of 19,200 baud.

Figure 15–10 illustrates a typical connector used in the transmission of RS–232 signals. The connector shown is a DB25 con-

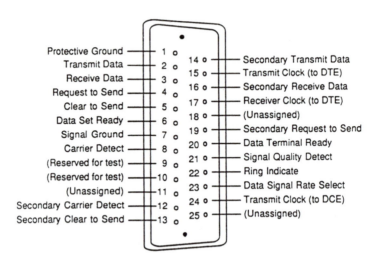

Figure 15–10   RS–232 DB25 Connector with Pinout Identification

nector. Notice that the connector is shaped like a D and that 25 pins are located inside it. All of the pins that are assigned are identified, as are the signal names found on these pins.

TRANSMISSION MODES

Computer-controlled devices that are connected together must communicate through wires or cables. As just discussed, one standard employed for this communication is the RS–232 standard. Several methods in which the communication can be established over the communication medium are discussed next. They are the simplex method, the half-duplex method, and the full-duplex method.

Simplex

Figure 15–11 illustrates the **simplex** method of transmission. This communication is established only between a computer and a printer. No communication is established between the printer and a computer terminal. Simplex transmission is the simplest of com-

Figure 15–11   Simplex Transmission

Figure 15–12   Half-Duplex Transmission

Figure 15–13   Full-Duplex Transmission

munication forms and is not often employed in industrial communication.

Half-Duplex

Figure 15–12 illustrates the **half-duplex** method of transmission. In this arrangement, the communication is established in both directions, but not simultaneously. Consequently, one device transmits while the other receives, and then they switch and transmit in the opposite direction.

Full-Duplex

Figure 15–13 illustrates the **full-duplex** method of transmission used in computer-controlled devices. Notice that messages can be sent and received at the same time. An example of this type of communication occurs when a robot controller commands a conveyor to move and a return signal from the conveyor to the robot controller indicates the conveyor is moving.

≡ NETWORKING THE COMMUNICATION

One of the most important areas for proper communication is the development of a network of the various components together. **Network communication** is a method of connecting the different devices together and allowing them to communicate with a central computer system or with one another.

There are several methods in which all the components can be connected together to form this networking arrangement. They are star topology, ring topology, or bus topology.

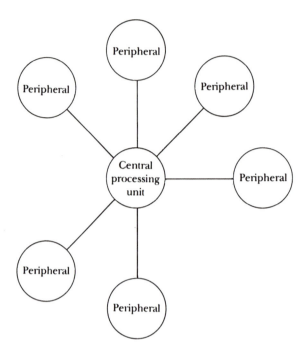

Figure 15–14   Star Topology

Star Topology

Figure 15–14 illustrates the **star topology** used in network communication. Notice that all of the peripheral components are connected by the network cable to some central processing point. All signals go through the central hub, no matter which two connecting devices are communicating.

As illustrated, the central processing unit has the peripheral devices connected in a star topology. The points that are connected from the central point are identified as *nodes* in the system.

There are two types of star connections, the active and the passive. A star connection is identified as active or passive if the center of the star has an active or a passive device used to interface the peripheral components.

The *active star topology* is illustrated in Figure 15–15. Here, the peripheral devices are connected to one central processing device, a mainframe computer that commands the direction of all signal processing. In this topology, the hub processor reads the command from a peripheral node and sends the information to another peripheral node that was addressed by the sending node. Notice that the central processing hub is the most important component in the entire system.

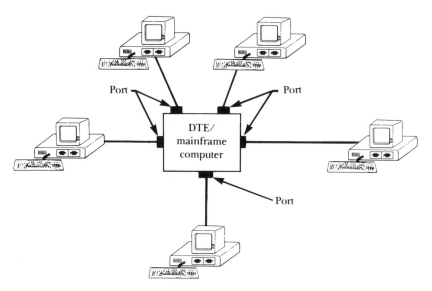

Figure 15–15 Active Star Topology

This type of connection has its strengths and its weaknesses. Since the hub controller and the data processing facility are the same, the same device does double the work in the system. It must maintain communication with all of the components within its topology and ensure that correct data is transferred between the various components within the topology. The weakness of this type of system is that when the hub processor breaks down, there is no communication at all within the system. The system will shut down until it is able to be repaired. This shutdown could, in turn, cause the assembly line to shut down, thus delaying production.

Figure 15–16 illustrates the *passive star topology* used in network communication. Notice here that the hub processing unit is a transformer. The transformer can only connect the different peripheral devices together so that they are able to communicate with one another. There is no processing power in the hub; all the processing power much be retained in the devices connected to the passive hub. In this type of topology, the devices connected to the hub must be intelligent enough to recognize when a message is addressed to them. That is, when a message is sent out from terminal 1 to terminal 2, for example, all other terminals connected to the topology will ignore the message.

Ring Topology

Figure 15–17 illustrates the **ring topology** used in many network connections. Notice in this type of topology that the nodes are con-

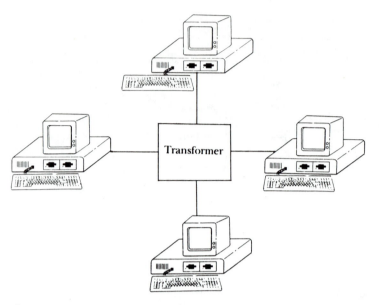

Figure 15–16   Passive Star Topology

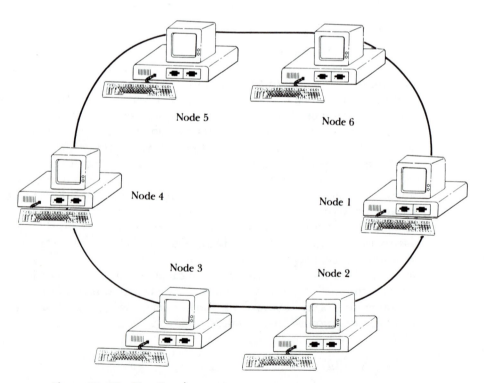

Figure 15–17   Ring Topology

nected together in series. When a message is generated in one terminal and is passed along to the different nodes within the ring topology, the devices must be able to identify the signal meant for them.

One advantage of this type of system is that all of the nodes are connected on one cable until the topology of the ring is formed. This topology allows for a signal to be passed from node 1 to node 3, the message must still pass through node 2. This means that the message is addressed so that only node 3 can read it. When node 3 takes the message off the line and reads it, a return message is placed back on the network. The return message passes from node 3 back to node 1.

There is one disadvantage to ring topology, however. As messages are placed on and off the network, errors may be generated, thus causing problems within the system.

Bus Topology

Figure 15–18 illustrates the **bus topology** used in many communication systems on the plant floor. In this type of system, a central cable is used for the main bus. From the main bus, many nodes can be connected. One reason why this type of topology is preferred over all the others is that it has only one central network cable for the main bus, and the nodes are connected directly through taps or modems into the bus network. Thus, the system can be modified easily if expansion or reduction of the system is required.

The communication method through this type of topology is very simple. When node 1 wants to communicate with node 3, it sends an addressing message for node 3 over the network. All of the nodes are listening to the network for their address. When node 3 reads its address, then it will receive the message and place an acknowledgment on the line for node 1. This acknowledgment will identify that the message was received or that the message required retransmission because it was not readable.

One of the major drawbacks to bus topology is that when a node breaks down, it will generally start to send out error messages. These error messages then fill the line with data that is trying to be read by the other devices connected to the bus network. The flood of error data transmitted by the faulty node does not allow for any other data to be transmitted over the bus. The problem is known as *streaming*. Another problem with bus topology is that when the bus breaks down, there is no communication within the entire system.

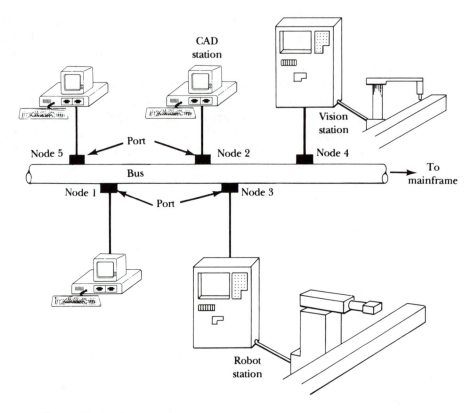

Figure 15–18   Bus Topology

NETWORK CABLES

There are several types of medium used to connect networks together. They are coaxial cable, twisted-pair wire, or fiber optics.
The choice of the medium to use in a network depends upon the
length of the medium run, the cost of the system, and the environment in which the medium will be used.

Coaxial Cable

Coaxial cable is the most popular type of medium to use when connecting a network together. The coaxial cable used is the same type
of cable used in the cable television industry to connect the various
cable locations together. The cable is constructed with a central
insulated conductor and a foiled shield around the insulated body.
The foiled, braided outside shield is used to keep unwanted signals

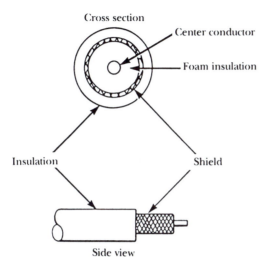

Figure 15–19 Coaxial Cable

from reaching the central conductor. Thus, a minimum number of error signals are generated by noise from within the factory—for example, noise generated from a motor starting or from a welding machine arc.

Besides reducing error signals, coaxial cable is used because it can transmit signals at a high rate of speed. The typical range of transmission on these lines ranges from 1 to 15 megabits per second (Mbps). Coaxial cable used in a factory networking system is very expensive to install, but can develop high-speed communication. Figure 15–19 shows the construction of coaxial cable in cross-sectional and side views.

Twisted-Pair Wire

The **twisted-pair wire** used in communication is basic telephone wire. The wire is constructed by twisting two insulated wires around each other. The insulation around each wire provides some insulation from electrical noise. Twisted-pair wire when used to connect the nodes together in high electrical noise areas will introduce induced errors into the cable. These errors will then flow through the topology and create errors in transmitted messages.

Speeds for twisted-pair connections are much slower than for coaxial cable connections. Typically, communication on these networks is only 300 to 9600 bits per second, thus ensuring reliable

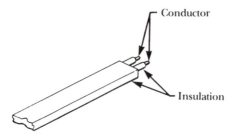

Figure 15–20   Twisted-Pair Wire

communication. Figure 15–20 illustrates the construction of a twisted pair of insulated wires used for a network connection.

Fiber Optics

One of the newest mediums used in the connections within networks is **fiber optics**. Fiber optics allow for the signal to be transmitted over light waves. The signal is converted to light energy and transmitted through a glass tube to the receiver. This medium is currently being used in telecommunication and has not been used in factory communication of networks because of the high cost of the cable and of its installation.

NETWORK-TO-NETWORK COMMUNICATION

As the use of computers increases within factory operation, it is becoming very important that computer-controlled areas are able to talk with one another. This network-to-network communication enables the different cells to collect data, pass information, and coordinate the manufacturing operation. In order to have this communication between the various networks within the manufacturing facility, each network must have the same computer protocol. **Protocol** is defined as an agreed format for communication between two or more parties. If this protocol is not established, then the networks within a factory are unable to communicate with one another, producing islands of automation within the computer-controlled factory.

Protocol Hierarchy

In order that communication can be established between different systems, a standard protocol has been established within the computer industry. This protocol has been identified by the Interna-

ISO Seven Levels
[7]Application level
[6]Presentation level
[5]Session level
[4]Transportation level
[3]Network level
[2]Link level
[1]Physical circuit level

Figure 15–21 Seven Levels for ISO Standard

tional Standard Organization, or **ISO** to be the framework for communication between networks. Figure 15–21 illustrates in block diagram form the seven levels of communication required to meet the network communication standard. They are as follows:

1. Physical circuit level
2. Link level
3. Network level
4. Transportation level
5. Session level
6. Presentation level
7. Application level

Level 1 in the ISO model identifies the physical circuits of the communication established between networks. This level specifies the electrical signals and voltages as well as the shape and pin assignments that DTE requires to communicate within the network. The level 1 operation refers to the type of hardware required. This hardware is generally off-the-shelf equipment that can be purchased. This equipment enables a data communication of bits of information to be transferred between the various nodes within the network.

The main function of level 2 is to control the physical link between the transmitting and receiving ends of level 1. This control ensures reliable transfer of information between DCE and DTE within the network.

The work of level 3 is to carry the message from one network node to another. The protocol used within this level enables a host DTE to communicate with another host DTE within the overall computer communication scheme.

Level 4, identified as the transportation level, is responsible for the safe transportation of the message or data stream from one application to another. Level 5, the session level, is responsible for management of the resources used in the network application. This management covers the operating system, the use of buffers, and the communication interface between networks.

Level 6, the presentation level, is responsible for mapping user requests into the network—for example, requests as to how terminals within the cell will be able to communicate with the network or how files will be transferred from one station to the next.

The final level of the ISO model, level 7, is the application level. This level is responsible at the host level for terminal or human operator interface with the system. The system is designed for human interaction with a terminal to call up various screens and menus for inputting and/or collecting data.

Connecting Networks

In many cases, the transfer of information will adhere to the ISO seven-level standards. But sometimes, the protocols between two networks will be different. If this is the case, then the message cannot be transferred between the two.

Figure 15–22 shows in block diagram form a device used to connect two networks together. This device is called a **gateway computer**. It is responsible for stripping away the protocol from network 1 and leaving just the address and message information. Once this is stripped away, then the protocol for network 2 is added. The message cannot be transferred into the next network for delivery.

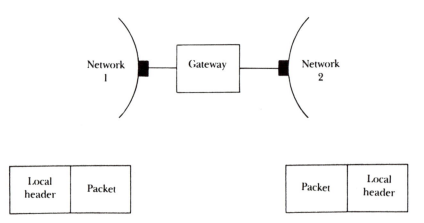

Figure 15–22 Block Diagram of Gateway Computer

≡ SUMMARY

Data communication can take on many forms. One method is the use of transmitting data via the voice. Another method is the electronic transmitting of digital-encoded signals.

Basically, digital signals are in one of two states, either on or off. These states are referred to as a logic 1 for the ON condition and a logic 0 for the OFF condition.

A standard for transmission of digital signals has been established through the American Standard Code for Information Interchange, or ASCII. This standard represents 128 characters of numbers, letters, and symbols converted to digital code.

Data terminal equipment (DTE) is used as the front end and as the reception end of digital transmissions. DTE includes mainframes, robot controllers, and minicomputers. Data communication equipment (DCE)—that is, a modem—is used to prepare and send digital signals over the transmission medium.

Digital signals travel at certain speeds over transmission mediums. These speeds are measured by bit rate or baud rate. Bit rate is the number of bits that can be transmitted per second, while baud rate is the number of times per second the signal changes logic states.

Each bit of data is identified either as a logic 1 or a logic 0. To save space in the transmission of data, bits are arranged into a pattern of 8 bits, or a byte of information. Bytes can be transmitted in either a serial or a parallel mode. Serial transmission requires only one wire to transmit data; parallel transmission requires 9 wires in a cable.

The RS–232 standard is used for the serial connection of computer-controlled devices. It identifies the type of connector, the number of pins, and the signal names for each of the pins.

The transmission of signals between computer-controlled equipment is accomplished through several different methods. They are the simplex method, the half-duplex method, and the full-duplex method.

The various transmission methods can be connected into networks to expand communication capabilities. These networks include star topology (active and passive), ring topology, and bus topology.

The topologies are connected by various mediums. The most common mediums used in the computer communication area are coaxial cable (the most popular type), followed by twisted-pair cable, and fiber optic cable (a newer technology).

Protocols are used in order to allow communication between the various networks connected within a plant. These protocols are used to establish a standard so that all of the communicating computer-controlled devices are based on the same foundation.

If two networks do not have the same protocol, they cannot communicate with each other. In order for them to communicate, a special computer called a gateway is employed. The gateway allows one protocol to be stripped away and a new protocol attached so that messages can be transmitted between networks.

KEY TERMS

ASCII
baud rate
bit rate
bits
bus topology
byte
coaxial cable
communication network
data communication
digital signal
fiber optics
full-duplex
gateway computer

half-duplex
ISO
logic 0
logic 1
network communication
parallel transmission
protocol
ring topology
serial transmission
simplex
strobe
star topology
twisted-pair wire

QUESTIONS

1. Give one application for a short-haul modem.

2. Name the two states of a digital signal.

3. How many bits are contained in a byte of information?

4. How many total characters are found in the ASCII code?

5. What is another name for DCE?

6. When data is transferred with a clocking pulse, the signal is said to be _____.

7. Which type of communication requires 9 lines to transfer data?

8. Name the first bit and the last bit transmitted in a serial transmission of data.

9. What is the bit rate for short-haul modems?

10. Identify the maximum distance that RS–232 signals can be transmitted without loss of signal.

11. The RS–232 standard specifies a maximum baud rate of _____.

12. Give an example of simplex transmission.

13. Name the type of transmission mode that can transmit in both directions, but cannot transmit simultaneously.

14. How many pins are found on a DB25 connector?

15. Identify the signal found on each of the following pins for a DB25 connector: pin 2, pin 6, pin 13, and pin 24.

16. The network topology that contains a transformer at the hub is identified as _____ topology.

17. A central cable carrying data with nodes connected to this cable would describe _____.

18. Coaxial cable can handle a maximum baud rate of _____.

19. What is the seventh level of the ISO standard protocol for computer-controlled communication?

20. What is the number of cables required to support a full-duplex connection?

21. Define *baud rate*.

22. State the advantage of serial communication as opposed to parallel communication.

23. What is the main function of the shield used in coaxial cable?

24. Give another name for a DTE device.

Glossary

This glossary provides the latest mechanical, electrical, and computer terms used in the robotics industry. Terms used throughout the text are included in this list.

— **AC input module:** I/O module that converts the various AC signals from user switches to the appropriate logic levels for use within the processor. An input module has two channels to accept two inputs from the peripheral devices.

— **AC output module:** I/O module that converts the various logic level signals to AC level signals to operate the various switches contained in the peripheral equipment.

— **acceleration:** The change in velocity as a function of time.

— **accuracy:** The capability of the manipulator to position the end effector at a specific predetermined location in space upon getting instructions from the controller.

— **ALU (arithmetic logic unit):** The unit of a computing system that performs logic and mathematical functions.

— **algorithm:** In computer or robotic software, a set of well-defined rules or procedures based on mathematical and geometric formulas for solving problems with a finite number of steps in the problem.

— **AML (a manufacturing language):** An IBM-developed robotics language used to program robotic movements.

— **analog:** Applied to electrical or computer systems, the term denotes the capability to represent data at continuously varying physical levels.

— **articulated arm:** A robot arm constructed to simulate the human arm. It consists of a series of rotary axes and joints, each powered by a servomechanism.

— **artificial intelligence (AI):** The capability of a computer to perform the functions that generally are performed by a human intelligence (e.g., learning, adapting, recognizing, classifying self-correction and/or improvements).

— **ASCII (American Standard Code for Information Interchange):** An industry-standard character set widely used for information interchange between data processing equipment and various associated equipment.

— **AS/RS (automatic storage/retrieval system):** The storage areas controlled by computer to automatically store and retrieve components for assembly.

— **assembly robot:** A mechanical manipulator used for putting together parts into subassemblies or complete products. Assembly robots are programmed to perform this assembly function.

— **automated guided vehicles system (AGVS):** Vehicles that are unmanned, computer-controlled, and preprogrammed to follow a specific path within a facility. Paths are established through buried wires in the floor. These vehicles are used to transfer parts to various workstations within the manufacturing facility.

— **asynchronous transmission:** Digital data signals formatted to have start and stop bits that identify the beginning and end of a character.

— **axis:** (1) A travelled path in space; a rotary or transactional joint. (2) A rotary or transactional joint in the robot. Identified by degrees of freedom.

— **backlash:** Free play in a power transmission system (e.g., a gear train).

— **backplane:** A printed circuit board located in the back of the controller used to connect other boards for communication of data.

— **base:** The platform or structure to which the robot's arm is attached.

— **baud rate:** A measure of the speed of data flow between the CPU and the peripheral devices it services. The term *baud* refers to the number of times the line condition changes in a second. It can be measured in signal events (bits) per second.

— **bit:** The smallest unit of information that can be stored and processed by a processor. A bit may assume only one of two conditions, either ON or OFF. Bits are organized into larger groups of words identified as bytes. Computer-controlled devices are often categorized by word size in bits (e.g., 16-bit computer, 32-bit computer). The number of bits in a word reflects the processing power of the computer-controlled device.

— **bus:** (1) In electrical design, a wire or path that power is allowed to flow through between two connection points. (2) In computer hardware, a circuit or group of circuits that provide communication between the CPU and the peripheral devices it supports.

— **byte:** A sequence of adjacent bits, usually eight, representing a character that is operated on as a whole.

— **CAD (computer-aided design):** The use of the computer to aid in the development of complex schematics or blueprints.

— **CAD/CAM (computer-aided design/computer-aided manufacturing): The integration of the computer to aid in the entire design-to-fabrication process of a product.

— **CAM (computer-aided manufacturing):** The use of the computer and digital technology to generate, collect, and distribute manufacturing data.

— **CCD camera:** A solid-state camera used to convert light images into digital signals through the use of charged coupled devices (CCD).

— **chain drive:** The method of transferring power from an actuator to a remote mechanism by means of a flexible chain and mating toothed sprocket wheel.

— **chip:** A small piece of silicon impregnated with impurities from transistors, diodes, and resistors. This combination forms the path for electrical current.

— **CIM (computer-integrated manufacturing):** The process of controlling the entire manufacturing process with the support of a central computing system. This system will collect, generate, store, and transfer data essential to the manufacturing, parts ordering, engineering, and sales process.

— **clock:** A device that initiates pulses for the synchronization of a computer operation.

— **closed loop:** A process of feeding back information and comparing that information with the original command signal.

— **computer program:** A set of instructions that are used to achieve a desired result. Often called a software program or package.

— **continuous path motion:** A robot program that allows the manipulator to follow a path on a constant timebase during teaching so that every point along the path of motion is recorded for future playback.

— **CPU (central processing unit):** That part of the computer or programmable controller that is used to carry out the instructions of the program.

— **cycle time:** The period of time from the start of one robot operation to the start of the next robot operation.

— **data communications:** The communication established between various pieces of equipment (e.g., between a CAD station and the CPU or between a robot controller and a programmable controller).

— **DCE (data communication equipment):** The equipment used to establish, maintain, and terminate communication between computer-controlled equipment and telephone lines.

— **digital-to-analog converter:** A device used to convert digital signals to analog signals.

— **duty cycle:** The time at which a device or a system will be active or at full power.

— **EIA (Electronics Industry Association):** The organization that sets standards in the electronics industry for the interfacing of equipment.

— **emergency stop:** A method that overrides all robot controls by removing power from the axis actuators, which brings the mechanical unit to a halt.

— **encoder:** A device used to convert linear or angular rotation into digital information.

— **end effector:** An actuator, gripper, or a driven mechanical device connected to the mechanical manipulator to perform work. This device is also referred to as end-of-arm tooling (EOAT).

— **envelope:** The entire area in which a mechanical manipulator can reach up, down, out, and side to side. See *work envelope*.

— **error signal:** The difference between a desired signal and the signal received.

— **expert system:** An intelligent computer program that uses knowledge and inference procedures to solve problems that require significant human expertise for their solution.

— **family of parts:** Parts that have the same geometric characteristics but that are different in physical measurement.

— **feedback loop:** The process and components used to take part of the output and return it to the input.

— **feedback devices:** Sensors used to collect the position of the device and transmit this information back to the robot controller.

— **FMS (flexible manufacturing system):** An arrangement of numerical-controlled machines (which are easily retooled), robots, and conveyors used to perform assembly operations.

— **flowchart:** A graphical representation of a sequence of operations that will be performed by the computer, robot controller, or manufacturing operation.

— **following error:** The input shaft of a servomechanism will rotate at a certain speed; the output shaft will rotate at the same speed but lag behind with an angle just sufficient enough to produce an error signal necessary to maintain the drive.

— **friction:** The resistance generated by the rubbing together of two objects.

— **full duplex:** The ability to send and receive data at the same time.

— **gain:** The increase in signal as it passes through a device. This term is often used in conjunction with amplification.

— **gantry robot:** A steel structure of legs and a bridge that suspends a mechanical manipulator. It is generally used to increase the work envelope of the robot.

— **gray scale:** A scale associated with vision systems that describes the different levels of contrast of objects. Many vision systems will have 256 levels of gray scale programmed in them.

— **GPM:** Gallons per minute.

— **gripper:** A device used to grasp, hold, transport, and deposit the component being held. Often referred to as the robot "hand."

— **ground:** A conduction between an electrical circuit or equipment chassis and the earth ground.

— **group technology:** The combination of family of parts together for a manufacturing operation. Also refers to the combination of resources for system implementation.

— **guard:** The safety feature used to protect people from danger in robotic systems.

— **half-bridge robot:** A Cartesian robot in which there is a north–south axis and an up–down axis but no east–west motion.

— **hardware:** The mechanical, electrical, and electronic devices that make up a programmable controller/computer/robot and the application components.

— **hunting:** The process that occurs when a control system's output continuously looks for its programmed position (e.g., when servo systems look for positional data). Rapid hunting turns into oscillation.

— **IEEE 488 Standard:** A digital data communications standard used in instrumentation electronics.

— **IEEE 802 Standard:** A protocol standard for local area network communication.

— **industrial robot:** A robot used for handling, processing, assembling, or inspecting materials or parts in a manufacturing operation.

— **input/output (I/O):** Communication that is common to the input and output of a computer-controlled device.

— **interfacing:** The method in which a robot communicates with peripheral devices connected within its cell. Sharing a common boundary.

— **interlock:** A condition established for safety on robots and other machining devices that will not allow operation until certain safety conditions are met.

— **islands of automation:** Stand-alone devices (e.g., robots, CAD,

CAM, and CNC) not connected together to form a total operating system.

— **joint:** The axis of a mechanical manipulator that will move when commanded.

— **KAREL:** A high-level user friendly programming language developed by GMF Robotics used to program robots and vision systems.

— **kinematics:** The plotting or animation of the movement of the axis of the manipulator through its work envelope. This action of the movement of the manipulator is generally accomplished through CAD simulations.

— **ladder diagrams:** Diagrams representing the operation of electrical circuits that are drawn between two vertical lines that represent the power lines.

— **laser:** A means of light amplification by stimulated emission of radiation. A laser can be used for the guiding of robotic welding applications.

— **lead-through:** A process of teaching the robot's position by leading the end-of-arm tooling through the path of the program.

— **LED (light-emitting diode):** A device used to convert electrical energy into light energy. LEDs are most often found in areas of robotic controllers for troubleshooting or indicating I/O functioning.

— **limit switch:** The switch used on robotic systems to identify when an action has occurred in the work cell. Limit switches are generally connected to the I/O of the controller.

— **load:** The external force applied to a body, or the energy required, or the act of applying force or requiring such energy.

— **local area network (LAN):** A facility that is connected together by cables to transmit information from one location to another. Generally, LAN is used to connect robots, CAD/CAM operations, and manufacturing processes together.

— **logic levels:** Associated with either highs or lows of bits of information.

— **machine tool:** A power-driven machine that is not hand held and that is used to cut, form, or shape metal.

— **manipulator:** The mechanical linkage or axis of a robot or a machine tool used to move parts from one location to another.

— **material handling:** A system comprised of conveyors, automatic retrieval and storage systems, lift trucks, automatic guided vehicles, computerized inventory, and robots for transferring parts for one operation to another.

— **material-handling robot:** The mechanical manipulator used to transfer parts from one operation to another operation within a material-handling system.

— **mean-time-between-failure (MTBF):** The average time a device will operate before it fails.

— **mean-time-to-repair (MTTR):** The average time required to repair a device once it fails.

— **MODEM (modulator/demodulator):** An electronic device used to transmit and receive electronic signals over telecommunication wires.

— **motor controller:** A device used to control the amount of power delivered to a motor.

— **NEMA standard:** An industry standard that identifies electric equipment approved by the National Electric Manufacturers Association (NEMA).

— **noise:** Extraneous signals in an electric circuit capable of causing interference with the desired signal.

— **numerical control (NC):** A technique of operating machine tools through the use of coded programs that automatically move the machine tool to a programmed location.

— **off-line programming:** The method used to program robots in which the programming device is not under the control of the CPU or the robot controller.

— **on-line programming:** The method used to program robots that is directly in control of that machine or robot operation. The programming is accomplished in real time.

— **optic sensor:** A device that converts light energy into electrical energy.

— **optical encoder:** Light source, chopper wheel, and photodiode used to identify the speed of an object.

— **output devices:** Devices such as solenoids, motors, starters, switches, or relays that receive data from the robot controller. These devices can also be used to input information to a controller.

— **pick-and-place robot:** A simple low-technology robot with two to four axes of motion used to move components from one location to another.

— **pinch point:** The area where a human can be trapped within the robot's work envelope without an avenue of escape.

— **pitch:** The up-and-down motion at an axis.

— **planetary drive:** A gear reduction arrangement that contains a spur gear, two or more planetary spur gears, and an internal toothed ring gear.

— **point-to-point control:** Robot motion that has only a specific amount of points along the path to which the manipulator travels.

— **polar motion:** Robot axes that have control and that form a spherical-shaped work envelope.

— **position error:** The difference between the commanded position of the servomechanism and the actual position obtained.

— **positioning accuracy and repeatability:** The accuracy of the robot is the measurement of the robot's ability to move to a programmed position. The repeatability is its ability to do a task over and over again.

— **PLC (programmed logic controller):** A stored programmed device used to replace relay ladder logic that controls the sequencing, timing, and counting of discrete events that happen in a manufacturing operation.

— **programming:** Industrial robots can be programmed in four basic methods: (1) lead-through, in which the robot is placed in the teach mode, the robot is moved through the various points in space, and each point is recorded for playback at a later time; (2) walk-through, in which the robot is placed in the teach mode, the manipulator is walked through the various points of the program, and each point is recorded for later playback during operation; (3) plug-in, in which a prerecorded program is uploaded from a magnetic tape, magnetic storage cassette, or floppy disk; (4) computer programming, in which programs that are written off-line on a computer are transferred to the robot's controller.

— **protocol:** A predefined set of rules under which the robot CPU operates.

— **real time:** The process operating on a CAD system that collects feedback to support the completion of the operation.

— **redundancy:** The process of building into a system parallel operations so that, in the event of a failure, the other system can complete the task.

— **repeatability:** The ability of the manipulator to position the end-of-arm tooling to the same programmed position under the same conditions.

— **resolver:** An encoder system that converts rotary or linear motion into angular positional data.

— **robot:** A reprogrammable multifunctional manipulator designed to move materials, parts, tools, or specialized devices through variable programmed motions for the performance of various tasks.

— **robotic cell:** One or two robots working together performing various tasks on a group of parts that may also require the use of other nonrobotic equipment.

— **roll:** The circular motion at an axis.

— **RS–232 Standard:** A communication standard for connecting peripheral equipment to CPU devices.

— **SCARA (selective compliance assembly robot arm):** A low-cost, high-speed robot arm used in the assembly operation. This robotic arm generally moves in a horizontal plane.

— **serial port:** An input/output device on a robot controller in which information is received and transmitted one bit at a time.

— **seam tracking:** The use of a mechanical probe with sensors feeding back detection of changes in the taught path of the end-of-arm tooling. This type of feedback system is generally found in arc welding or sealing applications.

— **servo-controlled robot:** A robot driven by a servomechanism in which feedback positional data is compared to command data, enabling the robot to move through a variety of points within the work envelope.

— **speed:** The maximum speed at which the tip of the robot can move through its work envelope.

— **stepping motor:** An electric motor whose windings are arranged in such a fashion that it allows the motor to rotate a minimum of 1/200 of a revolution upon command from the drive circuitry.

— **stop:** A mechanical device placed in the path of the robot's axes motion that will stop the motion of the robot axes or limit its travel.

— **tactile sensor:** A sensor required to make physical contact with a device in order to identify its location.

— **teach pendant:** The box connected to the controller that allows for the manual control of the motion of the robot axes.

— **tool centerpoint (TCP):** The programmed position of the center of the tool, such as grippers or a spot welding gun, connected to the end of the robot's arm.

— **torque:** The force tending to cause rotation of a body.

— **transducers:** Physical devices that are used to convert light, pressure, temperature, and other measurements into electrical signals.

— **turnkey system:** A robotic system that is totally the supplier's/vendor's responsibility to build, test, debug, and program, as well as train the customer's personnel on its operation.

— **velocity:** The measurement of the rate of change in motion.

— **vision system:** A device that converts light into digital signals that are used to identify components by shape and size. Vision systems are used to feed back information to a robotic system for the purpose of measuring offset positional data or the recognizing of parts.

— **work envelope:** The maximum reach of a robot while in full extension of the axes of the mechanical manipulator.

— **wrist:** Part of the manipulator arm joint that provides additional axes motion.

— **yaw:** The angular displacement of a moving body about an axis perpendicular to the line of motion and to the top side of the body.

Index